Jovan M. Nahman

Dependability of Engineering Systems

Springer

Berlin
Heidelberg
New York
Barcelona
Hong Kong
London
Milan
Paris
Tokyo

Engineering

ONLINE LIBRARY

http://www.springer.de/engine/

Jovan M. Nahman

Dependability
of Engineering Systems

Modeling and Evaluation

With 85 Figures

 Springer

Professor Dr. Jovan M. Nahman
University of Belgrade
Faculty of Electrical Engineering
Bul. K. Aleksandra 73
11120 Belgrade
Yugoslavia

Cataloging-in-Publication Data applied for

Die Deutsche Bibliothek - CIP-Einheitsaufnahme
Nahman, Jovan M:
Dependability of engineering systems: modeling and evaluation / Jovan M. Nahman.
Berlin; Heidelberg; NewYork; Barcelona; Hong Kong; London; Milano; Paris; Tokyo: Springer 2002
(Engineering online library)
ISBN 3-540-41437-1

ISBN 3-540-41437-1 Springer-Verlag Berlin Heidelberg New York

Springer-Verlag Berlin Heidelberg New York
a member of BertelsmannSpringer Science + Business Media GmbH

http://www.springer.de

© Springer-Verlag Berlin Heidelberg 2002
Printed in Germany

Typesetting: Camera ready copy from author
Cover-Design: medio Technologies AG, Berlin
Printed on acid-free paper   SPIN 10792277   62/3020 Rw 5 4 3 2 1 0

Preface

This book is intended to provide the interested reader with basic information on various issues of the dependability analysis and evaluation of engineering systems with the principal goal to help the reader perform such an analysis and evaluation. By the definition of the IEC International Standard 50(191) *dependability is the collective term used to describe the availability performance and its influencing factors: reliability performance, maintainability performance and maintenance support performance.* Dependability is a term used for a general description of system performance but not a quality which could be expressed by a single quantitative measure. There are several other quantitative terms, such as reliability, unreliability, time-specific and steady-state availability and unavailability, which together form a basis for evaluating the dependability of a system. A system is taken as dependable if it satisfies all requirements of the customers with regard to various dependability performances and indices. The dependability deals with failures, repairs, preventive maintenance as well as with costs associated with investment and service interruptions or mission failures. Therefore, it is a very important attribute of system quality.

The dependability evaluation is strongly based upon experience and statistical data on the behavior of a system and of its elements. Using past experience with the same or similar systems and elements, the prospective operation may be predicted and improved designs and constructions can be conceived. Hence, the dependability analysis makes it possible to learn from the past for better future solutions.

The first chapter of this book is devoted to the analysis of nonrenewable elements. The term *reliability* is defined as well as the associated indices. The methods of failure data analysis are outlined and illustrated by numerical examples. The duration of system (element) up state is a random variety which might have various cumulative probability distribution functions (Cdfs). For this reason, several typical Cdfs are described and characterized by relevant indices.

Renewable two-state systems (elements) are discussed in the second chapter to introduce the *availability* concept and the corresponding indices associated with renewal activities. Two-state systems with exponentially distributed up and down state residence times are analyzed as an introduction to *Markov systems*. Methods to determine the time-specific behavior of systems with generally distributed renewal times are described including the comparatively simple and efficient open loop sequence of states approach.

Markov models provide comprehensive means for the analysis of engineering systems with complex interactions of elements with one another and with the environment. They yield correct answers in the steady-state system analysis for many systems with nonexponentially distributed residence times. Furthermore, the approximation of the nonexponentially distributed states by sets of fictitious states with exponential distribution considerably widens the area of the applicability of Markov models. For this reason, the third chapter is devoted to a thorough presentation of these models in both time-specific and steady-state domains. Various aspects of model construction, simplification and calculation of relevant dependability indices are considered. Methods for the approximate determination of steady-state solutions based upon the transition matrix deviation concept and upon the matrix Gauss-Seidel approach are presented, enabling a fair solution for many practical systems. For simpler systems the solutions are obtainable in analytical form. These methods are applied to the dependability analysis of several characteristic examples including preventive maintenance, restricted repair, induced failures and cold standby systems.

Many engineering systems can be analyzed by means of adequate *network structures*. Among others, such systems are telecommunication and electric power networks, road traffic networks, water supply systems, etc. The fourth chapter deals in a detailed way with the methods of network dependability analysis. Together with the well known *minimum path* and *minimum cut* methods, some unconventional approaches are discussed enabling a proper modeling of the unreliable nodes and the dependence of branches. The possibilities of network partitioning and simplification are also mentioned.

The fifth chapter gives a brief overview of the dependability analysis methods based upon the inspection and evaluation of failure events. Presented are the *FMECA* and *FTA* approaches which are used for a qualitative evaluation of systems in various design phases. The operation of some engineering systems depends on the states of their components in such a complex manner that the impacts of various component states cannot be evaluated by a direct inspection. An acceptable method to analyze such systems is the *enumeration* of the set of events which might occur during system operation. For an effective and selective event set enumeration, a generalized concept of minimal cuts is introduced. The benefits which may be gained by the application of Kronecker algebra in state enumeration are also demonstrated.

The open loop *sequential approach* for transient dependability analysis described in the second chapter is extended, in the sixth chapter, to multi- state systems with generally distributed state residence times. We have successfully applied this method to finite term dependability prediction. It may be particularly useful in the analysis of mission time systems.

Stochastic simulation is an efficient method for the dependability analysis of many composite engineering systems. This is particularly the case for systems with complex interdependence of various events, systems with limited resources where the behavior in the past strongly affects the prospective system operation and, generally, for multi-state systems with nonexponential distributions of state residence times. The advantages and drawbacks of the simulation method when compared with the analytical methods and the limitations of these two approaches are discussed in the seventh chapter. Issues concerning the estimation of simulation errors, generation of random numbers and processing of results obtained are considered for both steady-state and time-specific system analysis.

As is known, dependability is inevitably associated with risks and costs for the investment in equipment and maintenance, on the one hand, and for the interruption of the service to the customers and system restoration, on the other hand. The eighth chapter briefly outlines a general concept of *system optimization* incorporating dependability and other performance requirements. In many cases the impacts of inadequate system service upon customers cannot be assigned a monetary value. In these cases the system dependability performance may be estimated from customer surveys and judgements expressed using linguistic attributes. The application of *fuzzy logic reasoning*, a new concept, may be helpful in these cases, which is demonstrated in this chapter. Finally, the ninth chapter deals with the *uncertainty* in dependability evaluation. Many input parameters used in dependency analysis suffer from uncertainty. That is especially the case when new equipment and/or constructions are designed and operated for which the relevant dependability data can be only roughly assessed. This chapter suggests how to handle uncertainties and how to assess their effects upon the results of the dependability analysis by applying *operations with fuzzy numbers*. This is also a new area of research which should be, in our opinion, implemented in the dependability analysis of practical systems. The results obtained in such a way are better attuned to reality than the results from the deterministic crisp approach. The latter appears to be only a particular case of the fuzzy analysis. It was the reason to include this material in the book.

The text is illustrated with more than 50 examples, for better understanding and clarity. At the end of each chapter there is a set of problems for self-study. A basic knowledge of probability theory, Laplace transforms, differential equations and matrix algebra is assumed as a prerequisite. Tables of characteristic probability distributions are not included as the corresponding data are easily generated using MATLAB or similar available software packages.

Belgrade, Summer 2001   Jovan M. Nahman

Contents

1 Nonrenewable Two-State Systems

Two-state systems dealt with in this chapter are systems which operate until a failure occurs interrupting their functioning. No renewals are considered. However, the theory developed for nonrenewable systems is completely applicable to renewable systems too, for the prediction of their behavior between two successive failures. The term *system* is used further on in this chapter for convenience. However, it also applies to components of a system individually. Many electronic components are typical examples of nonrenewable two-state systems.

1.1
Characteristic Functions and Indices

Nonrenewable systems are systems which cannot be repaired. This means that the life of such systems lasts as long as the time from their construction or installation until the first failure. The *reliability* of such systems is defined as the probability that the system will not fail before time t

$$R(t) = \Pr\{T > t\} \tag{1.1}$$

with T denoting the duration of the sound state of the system and t being the time elapsed from the construction or installation of the system. Hence, $R(t)$ is the probability that the system will be sound at time t. $R(t)$ is a dynamic index as it changes with time. It is clear from common sense that the reliability of any system tends to zero with time.

The *unreliability* of a system is defined as the probability that the system is down at time t. The failure can happen at any instant before or at t. Clearly, the corresponding mathematical definition is

$$Q(t) = \Pr\{T \leq t\} \tag{1.2}$$

From (1.2) it follows that $Q(t)$ is the *Cumulative Probability Distribution Function* (Cdf) of random variable T. Unreliability is a dynamic index too.

With regard to (1.1) and (1.2) it follows that

$$R(t) + Q(t) = 1 \tag{1.3}$$

The *Probability Density Function* (pdf) of random variable T is by definition

$$f(t) = \frac{dQ(t)}{dt} \qquad (1.4)$$

Using the arguments of probability theory we may write

$$dQ(t) = Q(t+dt) - Q(t) = \Pr\{t<T\leq t+dt\} \qquad (1.5)$$

From (1.4) and (1.5) it follows that $f(t)dt$ is the probability that the system will fail during time interval $(t, t+dt)$. Function $f(t)$ may be called the *time-specific failure frequency*.

By taking a derivative of (1.3) with respect to time we obtain, using (1.4)

$$\frac{dR(t)}{dt} = - f(t) \qquad (1.6)$$

According to (1.4)

$$Q(t) = \int_{-\infty}^{t} f(x)dx = \int_{0}^{t} f(x)dx \qquad (1.7)$$

Function $f(x)$ is zero for negative arguments if time x is counted from the instant of the construction or installation of the system.

With regard to (1.3) and (1.7) we have

$$R(t) = 1 - \int_{0}^{t} f(x)dx = \int_{0}^{\infty} f(x)dx - \int_{0}^{t} f(x)dx = \int_{t}^{\infty} f(x)d. \qquad (1.8)$$

Another important reliability and dependability index is the *failure transition rate* defined as

$$\lambda(t) = \lim_{\Delta t \to 0} \frac{\Pr\{t<T\leq t+\Delta t|(T>t)\}}{\Delta t} \qquad (1.9)$$

The numerator in (1.9) is the probability that the system will fail during time interval $(t, t+\Delta t)$ given that it has not failed before t. This relationship can be reformulated. The expression for the probability of the coincidence of two events, say A and B, is

$$\Pr\{AB\} = \Pr\{A|B\} \Pr\{B\} \qquad (1.10)$$

and hence,

$$\Pr\{A \mid B\} = \frac{\Pr\{AB\}}{\Pr\{B\}} \tag{1.11}$$

Having regard to (1.11) and (1.1) the conditional probability in (1.9) equals

$$\Pr\{t < T \le t + \Delta t) \mid (T > t)\} = \frac{\Pr\{(t < T \le t + \Delta t) \ (T > t)\}}{\Pr\{T > t\}} = \frac{\Pr\{t < T \le t + \Delta t\}}{R(t)} \tag{1.12}$$

The final form of (1.12) is simple because event B is included in event A.
    Bearing in mind (1.12), (1.4) and (1.5), expression (1.9) transforms into

$$\lambda(t) = \frac{f(t)}{R(t)} \tag{1.13}$$

If $f(t)$ is expressed in (1.13) with respect to (1.6), a differential equation is obtained for $R(t)$ whose solution is

$$R(t) = \exp\left(-\int_0^t \lambda(x)dx\right) \tag{1.14}$$

From the previous consideration it is clear that indices $R(t)$, $Q(t)$, $f(t)$ and $\lambda(t)$ determine one another. Therefore, a knowledge of any one of them enables the determination of the rest.
    A clear orientation for practical reliability assessment yields the *mean time to first failure* (MTFF) which is defined as the mathematical expectation of the system up time

$$m = E\{T\} = \int_0^\infty tf(t)dt = -\int_0^\infty tR'(t)dt \tag{1.15}$$

The lower limit of integrals in (1.15) is zero for the same reason as before. If partial integration is used to solve the second integral in (1.15), we obtain

$$m = -tR(t)\Big|_0^\infty + \int_0^\infty R(t)dt = \int_0^\infty R(t)dt \tag{1.16}$$

The first term in the first expression in (1.16) is zero as the rate of decrease of $R(t)$ is higher than the rate of increase of t for large t. Index m is a steady-state parameter as it does not depend on t.

All the indices considered so far may be defined and used for renewable systems, i.e. for systems which after failure are repaired or replased by another system of the same type. For such systems the previously considered indices are used to characterize the reliability of the system from the instant of (re)installation until the next failure. If the system is renewed by repair it is implied that it is repaired as new.

From the reliability indices defined previously, various important estimates of system behavior can be deduced. Several examples which follow demonstrate this.

The probability that the system will fail during time interval (t_1, t_2) can be calculated using any of the following expressions

$$\Pr\{t_1 < T \le t_2\} = \int_{t_1}^{t_2} f(x)dx = Q(t_2) - Q(t_1) = R(t_1) - R(t_2) \tag{1.17}$$

The probability that the system will be up during time interval (t_1, t_2) equals, with regard to (1.17)

$$1 - \Pr\{t_1 < T \le t_2\} = 1 - R(t_1) + R(t_2) = Q(t_1) + R(t_2) \tag{1.18}$$

The probability that the system will fail during time interval (t_1, t_2) if it has not failed before t_1 equals, with regard to (1.11) and (1.17)

$$\Pr\{(T \le t_2) | (T > t_1)\} = \frac{\Pr\{t_1 < T \le t_2\}}{\Pr\{T > t_1\}} = \frac{R(t_1) - R(t_2)}{R(t_1)} :$$
$$= 1 - \frac{R(t_2)}{R(t_1)} \tag{1.19}$$

As can be seen, the conditional failure probability (1.19) is greater than the unconditional probability given by (1.17) for all $t_1 > 0$ as then $R(t_1) < 1$. This result is logical as (1.17) implies that the system might fail before t_1 which decreases the probability of failure during interval (t_1, t_2).

The probability that the system will be up during time interval (t_1, t_2) given that it was sound before t_1, equals according to (1.19)

$$\Pr\{(T > t_2) | (T > t_1)\} = 1 - \Pr\{(T \le t_2) | (T > t_1)\} = \frac{R(t_2)}{R(t_1)} \tag{1.20}$$

The latter probability can simply be determined using the failure transition rate. If we express the reliability in (1.20) with regard to (1.14), the following expression is obtained after elementary manipulations

$$\Pr\{(T>t_2)|(T>t_1)\} \;=\; \exp\left(-\int_{t_1}^{t_2}\lambda(x)dx\right) \tag{1.21}$$

Example 1.1

Consider a system whose failure transition rate is assessed to be proportional to the in-service time: $\lambda(t) = at$ with $a = 1$ fl./(1000 h)^2. The probability that the system will be up after being 500 h in service is to be determined.

Taking into account (1.14) we may write

$$R(t) \;=\; \exp\left(-\int_0^t axdx\right) \;=\; \exp\left(-\frac{at^2}{2}\right)$$

$$R(500) \;=\; \exp\left(-10^{-6}\,\frac{1}{2}\,500^2\right) \;=\; 0.8825$$

The probability of system failure during 500 h is $1 - 0.8825 = 0.1175$.

$\square$

Example 1.2

Assume that the system from *Example 1.1* is operating for 300 h. The probability that the system will be sound for another 100 h is to be determined.

The probability reqired is, according to (1.20)

$$\frac{R(t_2)}{R(t_1)} \;=\; \exp\left(-\frac{a}{2}(t_2^2 - t_1^2)\right)$$

If we insert $t_1 = 300$ h and $t_2 = 300+100 = 400$ h into this expression we find that the required probability is 0.9656. The probability that the system will fail during time interval (t_1,t_2) is, according to (1.19), $1 - 0.9656 = 0.0344$.

$\square$

1.2
Analysis of Failure Data

The system reliability indices are evaluated by processing the data acquired during the exploitation of a sufficiently large number of systems of the same type operating under similar conditions. The systems are observed from the beginning of the exploitation until the occurrence of the first failure.

Let $n(t)$ be the number of systems which failed before t and $N(t)$ the number of systems that remain sound during time interval t. By $\Delta n(\Delta t_{12})$ we denote the number of systems that failed during time interval $\Delta t_{12} = t_2 - t_1$.

With regard to (1.2) the unreliability of a system at time instant t_1 is defined as the probability that the system will fail before or at this time instant. The aforementioned probability can be evaluated from the observation data as

$$Q(t_1) = \frac{n(t_1)}{N(0)} \tag{1.22}$$

The pdf of the sound operation duration during time interval Δt_{12} can be determined with regard to (1.4) and (1.22) by applying the expression

$$f(\Delta t_{12}) = \frac{\Delta n(\Delta t_{12})}{N(0)\Delta t_{12}} \tag{1.23}$$

The failure transition rate value during time interval Δt_{12} can be evaluated by bearing in mind (1.13) and (1.23) as

$$\lambda(\Delta t_{12}) = \frac{\Delta n(\Delta t_{12})}{N(t_1)\Delta t_{12}} \tag{1.24}$$

The results obtained using the above expressions are the more credible the larger is the number of the systems under observation ($N(0)$). Time intervals Δt_{12} are selected to encompass several failures of the systems. The whole time domain under consideration is usually subdivided into K intervals of the same duration Δt . The number K should be selected in such a way as to preserve the substantial information on the variation of reliability indices with time which could be distorted by an extensive averaging of data. By experience acquired from many practical applications it is suggested to determine K by applying the following empirical relationship [4]

$$K = 1 + 3.3 \log(n) \tag{1.25}$$

with n being the total number of observed failures.

Another empirical formula is [5]

$$K = \sqrt{n} \tag{1.26}$$

Example 1.3

For illustration, Table 1 gives the failure data of a hypothetical system and the $Q(t)$ and $\lambda(t)$ functions calculated from these data by applying (1.22) and (1.24). It is presumed that 40 systems of the same type have been observed from the beginning of their operation until the failure. Hence, in this case $n = N(0) = 40$.

Table 1.1. Failure data and calculated $Q(t)$ and $\lambda(t)$ functions

Time intervals Δt, days	$\Delta n(\Delta t)$	$Q(t)$	$\lambda(t)$ fl./(10^3 days)
0	0	0	-
1 – 200	4	4/40 = 0.100	4/(40×0.2) = 0.5
201 – 400	9	13/40 = 0.325	9/(36×0.2) = 1.25
401 – 600	8	21/40 = 0.525	8/(27×0.2) = 1.48
601 – 800	7	28/40 = 0.700	7/(19×0.2) = 1.84
801 – 1000	6	34/40 = 0.850	6/(12×0.2) = 2.50
1001 – 1200	6	40/40 = 1.000	6/(6×0.2) = 5.00

Functions $Q(t)$ and $R(t)$ are displayed in Fig. 1.1. $R(t)$ is calculated from $Q(t)$ using (1.3). The calculated failure transition rate in terms of time is depicted in Fig. 1.2.

The values from Table 1.1 are taken to relate to the midpoints of the corresponding time intervals. Linear interpolation between succesive points is applied.

If T_i is the observed time to failure of system i, the system expected time to failure is

$$m = \frac{1}{N(0)} \sum_{i=1}^{N(0)} T_i \tag{1.27}$$

All expressions quoted in this section may be conditionally applied for evaluating the reliability indices of a system by observing this system only. During a sufficiently long period of observation the system will fail several times. The periods of sound operation might be treated as independent samples of system operation and processed in the same way as if these samples were acquired from a set of systems. Clearly, such an approach is valid only if the system under consideration is repaired as new after each failure. If this is not the case the data for successive periods of operation would not be independent as the system would behave differently in various periods depending on its age and on the number and the quality of repairs in the past.

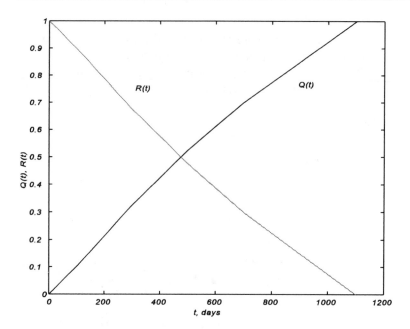

Fig. 1.1 Functions $Q(t)$ and $R(t)$ from data in Table 1.1

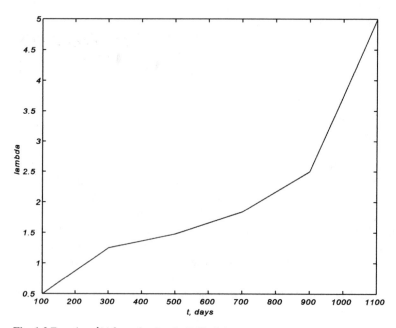

Fig. 1.2 Function $\lambda(t)$ from the data in Table 1.1

1.3
Characteristic Probability Distributions

For solving various dependability problems and drawing some general conclusions or estimates it is convenient to approximate as closely as possible the empirical $Q(t)$ function by a known theoretical Cdf. Then, various indices can simply be calculated from one another by applying elementary mathematical operations, as shown in Section 1.1. In this section we will describe a few theoretical probability distributions which can serve as good approximations of various distributions which might be deduced from empirical data [1–7].

Further on the following notation is used:

X - random variable

x - possible value of X

$F(x)$ - Cdf of X:

$$F(x) = \Pr\{X \le x\} \tag{1.28}$$

$f(x)$ - pdf of X:

$$f(x) = \frac{dF(x)}{dx} \tag{1.29}$$

M - mathematical expectation of X :

$$M = \mathrm{E}\{X\} = \int_{-\infty}^{\infty} x f(x) dx \tag{1.30}$$

$V(x)$ - X variance :

$$V(x) = \int_{-\infty}^{\infty} (x-M)^2 f(x) dx \tag{1.31}$$

σ - standard deviation of X :

$$\sigma = \sqrt{V(x)} \tag{1.32}$$

If X is a system state residence time, the lower bounds in (1.30) and (1.31) are zeros as X cannot take negative values.

1.3.1
Exponential distribution

The exponential Cdf is defined by the following relationship

$$F(x) = \begin{cases} 1-\exp(-\lambda x) & x \geq 0 \\ 0 & x > 0 \end{cases} \qquad (1.33)$$

From (1.33) it follows, on the basis of (1.29)

$$f(x) = \begin{cases} \lambda \exp(-\lambda x) & x > 0 \\ 0 & x \leq 0 \end{cases} \qquad (1.34)$$

Function $f(x)$ is displayed in Fig. 1.3 for two different values of λ.

With regard to the general definitions given at the beginning of this section it is found that

$$M = \frac{1}{\lambda} \qquad (1.35)$$

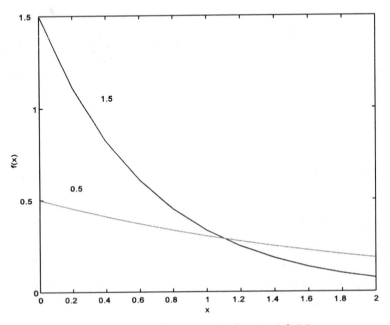

Fig. 1.3 Pdf of the exponential distribution for $\lambda=1.5$ and $\lambda=0.5$

$$V(x) = \frac{1}{\lambda^2} \tag{1.36}$$

$$\sigma = \frac{1}{\lambda} \tag{1.37}$$

If X is the up time of a nonrenewable system then M is the mean time to first failure, which follows from (1.15) and (1.30). As can be observed from (1.35) and (1.37), for the exponential Cdf parameters M and σ coincide.

From (1.34) and (1.13) we deduce that for the exponential Cdf

$$\lambda(x) = \lambda \tag{1.38}$$

This means that systems having the exponentially distributed up time fail with a constant rate in time.

The exponential Cdf possesses an important property which, as we will see in the next chapters, enables relatively simply modeling of multi-state systems with exponentially distributed state residence times. The probability that a system which has been up until time t_1 will fail during time interval $t_2 - t_1$ is for an exponential Cdf of system up time, with regard to (1.14) and (1.19)

$$\mathrm{Pr}\{(T \le t_2) | T > t_1)\} = 1 - \exp\left(-\int_{t_1}^{t_2} \lambda dx\right) = \tag{1.39}$$

$$= 1 - \exp(-\lambda(t_2 - t_1))$$

As may be observed, the analyzed probability does not depend upon the time t_1 spent in the state but only upon the time in the future as viewed from instant t_1. This means that the exponential distribution models systems lacking memory.

If Δt_{12} is small, the exponential function in (1.39) may be approximated by the first two terms of its power series expansion which yields

$$\mathrm{Pr}\{(T \le t_2) | (T > t_1)\} \approx \lambda(t_2 - t_1) \tag{1.40}$$

The exponential Cdf can successfully model the up times of many systems which are exposed to rare, random stresses often being of environmental origin. Many components of electronic and electrical systems behave very much in this way. In reality, however, the failure transition rate is not ideally constant during the whole lifetime of the aforementioned systems. The variation of the transition rate with time can be approximately represented by a bathtub curve as depicted in Fig. 1.4 [3,5]. Three characteristic time periods can be identified by inspection of Fig. 1.4. The first

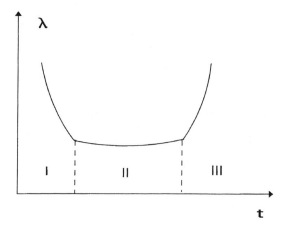

Fig. 1.4 Variation of failure transition rate in time

period is characterized by relatively high transition rates usually due to possible damage in transport and installation errors. The second period begins after the early failures have been cured and lasts until the beginning of the last period with gradually increasing failure rates due to deterioration of materials. The second period is usually the longest. The technical life of a system should be taken as terminated when the corrective maintenance and outage costs rise beyond certain acceptable limits.

1.3.2
Weibull Distribution

Weibull Cdf and pdf are described by the following expressions [5]

$$F(x) = \begin{cases} 1 - \exp\left(-(\frac{x-v}{\alpha})^{\beta}\right) & x \geq v \\ 0 & x < v \end{cases} \tag{1.41}$$

$$f(x) = \begin{cases} \frac{\beta}{\alpha}(\frac{x-v}{\alpha})^{\beta-1} \exp\left(-(\frac{x-v}{\alpha})^{\beta}\right) & x \geq v \\ 0 & x < v \end{cases} \tag{1.42}$$

As may be seen, a Weibull distribution is characterized by three parameters. Parameter $v (- \infty < v < \infty)$ controls the position of the pdf with respect to the ordinate axis and is usually called the location parameter. Parameter $\beta\,(\beta > 0)$ mainly affects the shape of the distribution while parameter $\alpha\,(\alpha > 0)$ determines it proportions. Therefore, these parameters are called the shape and scale parameters, respectively.

By applying the general definitions (1.30) and (1.31) to (1.41) and (1.42) we obtain

$$M = \alpha\Gamma(1+\frac{1}{\beta}) + v \tag{1.43}$$

$$V(x) = \alpha^2\left\{\Gamma(1+\frac{2}{\beta}) - \left[\Gamma(1+\frac{1}{\beta})\right]^2\right\} \tag{1.44}$$

with $\Gamma(x)$ being the Gamma function defined as

$$\Gamma(x) = \int_0^\infty y^{x-1}\exp(-y)dy \qquad x>0 \tag{1.45}$$

The partial integration of (1.45) yields [5]

$$\Gamma(x) = (x-1)\ \Gamma(x-1) \tag{1.46}$$

For positive integer x expression (1.45) converts into

$$\Gamma(x) = (x-1)! \tag{1.47}$$

As we see from (1.43), the mean time to failure M depends linearly upon parameter v. However, this parameter does not affect the variance $V(X)$, as can be observed from (1.44).

By applying the general expression (1.13) we obtain from (1.41) and (1.42)

$$\lambda(x) = \frac{\beta}{\alpha}(\frac{x-v}{\alpha})^{\beta-1} \tag{1.48}$$

It is notable from (1.48) that for $\beta > 1$ the failure rate for a Weibull distribution increases with x. If random variable X is the residence time in the operating system state, a Weibull distribution can model the gradual wearout of the system which is applicable to many mechanical assets.

For $\beta=1$ and $v=0$ a Weibull distribution converts into an exponential distribution which means that this distribution can be regarded as a special case of the Weibull distribution. For illustration, Fig. 1.5 displays a Weibull pdf for $v=0$ and $\alpha=1$ for

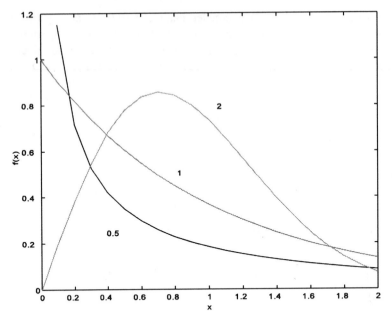

Fig. 1.5 Weibull pdf for $\nu=0$, $\alpha=1$ and various β

various values of β. As can be observed, parameter β substantially affects the shapes of the curves.

1.3.3
Uniform Distribution

The random variable X is uniformly distributed if its Cdf and pdf have the following forms

$$F(x) = \begin{cases} 0 & x \leq a \\ \dfrac{x-a}{b-a} & a < x \leq b \\ 1 & x > b \end{cases} \tag{1.49}$$

$$f(x) = \begin{cases} \dfrac{1}{b-a} & a < x \leq b \\ 0 & \text{otherwise} \end{cases} \tag{1.50}$$

These functions are displayed in Figs. 1.6 and 1.7.

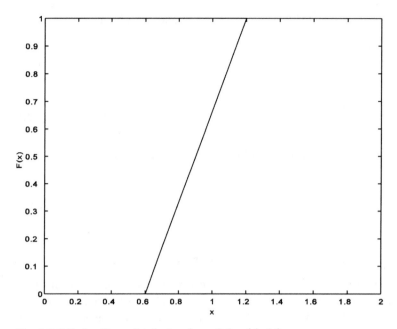

Fig. 1.6 Cdf of uniform distribution for a=0.6 and b=1.2

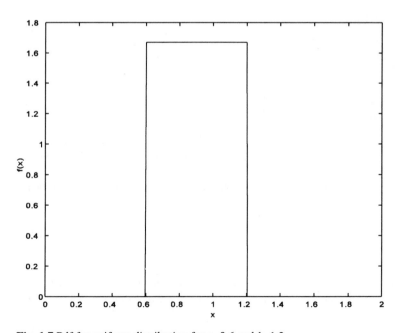

Fig. 1.7 Pdf for uniform distribution for a=0.6 and b=1.2

By applying (1.30) and (1.31) to uniform distribution we obtain

$$M = \frac{a+b}{2} \tag{1.51}$$

$$V(X) = \frac{(a-b)^2}{12} \tag{1.52}$$

The transition rate for a uniform distribution equals, with regard to (1.13)

$$\lambda(x) = \begin{cases} 0 & x \leq a \\ \dfrac{1}{b-x} & a<x<b \end{cases} \tag{1.53}$$

By definition, the probability that the uniformly distributed random variable X will be within a sub-interval of interval (a,b) is proportional to the length of this sub-interval. The narrower the interval (a,b) is, the less random is the random variable. The uniform distribution may be used to model the preventive maintenance duration in many cases.

1.3.4
Normal Distribution

The pdf of the normal distribution is described by the following expression

$$f(x) = \frac{1}{\alpha\sqrt{2\pi}} \exp\left(-\frac{(x-\mu)^2}{2\alpha^2} \right) \tag{1.54}$$

with μ ($-\infty<\mu<\infty$) being the location parameter and α ($\alpha^2>0$) representing both the shape and scale parameters.

The Cdf of the normal distribution is calculated by numerical integration of the pdf

$$F(x) = \int_{-\infty}^{x} f(u)du \tag{1.55}$$

as the solution of the integral in (1.54) cannot be obtained in analytical form.

By applying (1.30) and (1.31) we obtain for a normal distribution

$$M = \mu \tag{1.56}$$

$$V(x) = \alpha^2 \qquad (1.57)$$

$$\sigma = \alpha \qquad (1.58)$$

The pdf of the normal distribution is displayed in Fig.1.8 for $\mu = 1$ and two values of parameter α. As can be observed, the pdf is symmetrical about its mean value $x=M$, which means that random variable X may be greater or less than M with the same probability. It is notable that X may have negative values too. Because of the latter, the normal distribution cannot be used for modeling the system state residence times. However, a normal distribution describes well the random variables which are affected by a large number of various factors having small effects individually [1]. For this reason the normal distribution is often utilized to estimate the errors of prediction of various quantities which are not known with certainty.

In order to simplify the application of a normal distribution in practical calculations, the *standard normal distribution* is defined [9]. The pdf and Cdf of this distribution are

$$\phi(z) = \frac{1}{\sqrt{2\pi}} \exp\left(-\frac{z^2}{2}\right) \qquad (1.59)$$

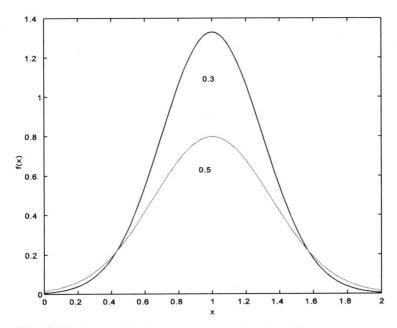

Fig. 1.8 Pdf of normal distribution for $\mu=1$, $\alpha=0.3$ and $\alpha=0.5$

$$\Phi(z) = \int_{-\infty}^{z} \phi(u)du = 0.5 + \int_{0}^{z} \phi(u)du \tag{1.60}$$

Specialized books provide tables for determining the values of $\Phi(z)$ [8].
The values of random variables Z and X are simply correlated

$$z = \frac{x-\mu}{\alpha} \tag{1.61}$$

Thus, with regard to (1.55), (1.60) and (1.61)

$$F(x) = \Phi(\frac{x-\mu}{\alpha}) \tag{1.62}$$

For modeling the system state residence states the *truncated normal distribution* may be applied defined as [8,10]

$$F(x) = \begin{cases} c\Phi(\frac{x-\mu}{\alpha}) - c\Phi(\frac{v-\mu}{\alpha}) & x \geq v \geq 0 \\ \\ 0 & x < v \end{cases} \tag{1.63}$$

where

$$c = \left(1 - \Phi(\frac{v-\mu}{\alpha})\right)^{-1} \tag{1.64}$$

Parameter c is introduced in order to satisfy the general property of a probability distribution: $\lim F(x) = 1$ for $x \to \infty$.

1.3.5
Gamma Distribution

The pdf of gamma distribution equals

$$f(x) = \begin{cases} \frac{\beta\Theta}{\Gamma(\beta)}(\beta\Theta x)^{\beta-1}\exp(-\beta\Theta x) & x > 0 \\ \\ 0 & x \leq 0 \end{cases} \tag{1.65}$$

The gamma distribution is characterized by the shape parameter β and by the scale parameter Θ. The Cdf of this distribution has to be calculated by numerical integration of the pdf as indicated by (1.55) because it cannot be expressed in closed analytical form. Regarding (1.30), (1.31) and (1.65) we obtain

$$M = \frac{1}{\Theta} \tag{1.66}$$

$$V(x) = \frac{1}{\beta\Theta^2} \tag{1.67}$$

For $\beta=1$ the gamma distribution converts into the exponential distribution. For large values of β the gamma distribution approaches the normal distribution. Fig. 1.9 displays the pdf of the gamma distribution.

If $\beta=k$, where k is a positive integer, the gamma distribution converts into *Erlang's distribution of order k* whose Cdf equals

$$F(x) = \begin{cases} 1-\exp(-k\Theta x)\sum_{i=0}^{k-1} \frac{(k\Theta x)^i}{i!} & x>0 \\ \\ 0 & x\leq 0 \end{cases} \tag{1.68}$$

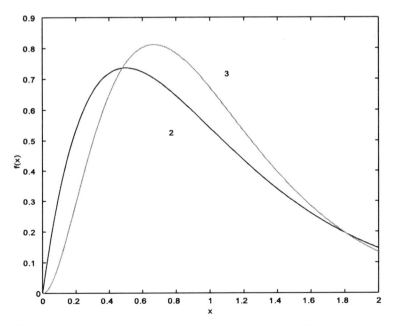

Fig. 1.9 The pdf of the gamma distribution for $\Theta=1$, $\beta=2$ and $\beta=3$

Erlang's distribution describes, for instance, the up time of a system composed of k components whose lifetimes are all exponentially distributed with a failure transition rate being equal to $k\Theta$. Only one component is operating at a time while other components are in the cold standby mode and cannot fail in this mode. When the unit in service fails it is immediately replaced by a standby unit, if any is left. It is clear that the up time of this system equals the sum of components up times.

1.3.6
Lognormal Distribution

The random variable X has the lognormal distribution if $\ln X$ is normally distributed. Consequently, the Cdf of the lognormal distribution is given by the expression

$$F(x) = \Phi(\frac{\ln x - \mu}{\alpha}) \qquad x \geq 0 \tag{1.69}$$

By differentiating (1.69) we obtain, bearing in mind (1.29)

$$f(x) = \frac{1}{x\alpha\sqrt{2\pi}} \exp\left(- \frac{(\ln x - \mu)^2}{2\alpha^2} \right) \tag{1.70}$$

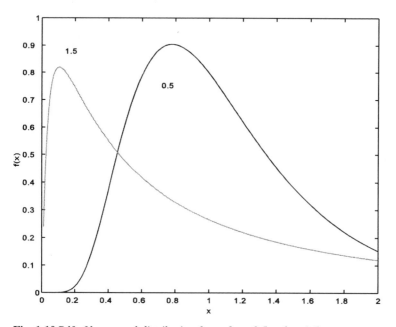

Fig. 1.10 Pdf of lognormal distribution for $\mu=0$, $\alpha=0.5$ and $\alpha=1.5$

From (1.30), (1.31) and (1.70) it follows that

$$M = \exp\left(\mu + \frac{\alpha^2}{2}\right) \tag{1.71}$$

$$V(x) = \exp\left(-(2\mu + 2\alpha^2)\right) - \exp(2\mu + \alpha^2) \tag{1.72}$$

It has been found that the repair duration of some assets may be successfully modeled by the lognormal distribution [3]. The lognormal pdf is depicted in Fig. 1.10, for illustration.

1.3.7
Combination of Distributions

If the statistical data do not fit sufficiently well to any of the theoretical distributions, a combination of various distributions may be used. The pdf of a combination of distributions is obtained as

$$f(x) = \sum_{i=1}^{n} a_i\, f_i(x) \qquad a_i > 0 \tag{1.73}$$

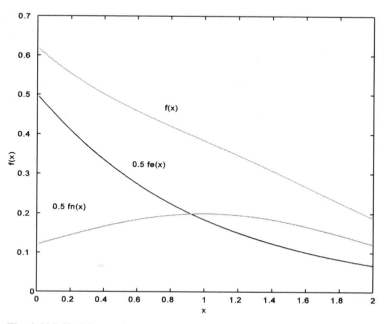

Fig. 1.11 Pdf of the combination of the exponential *(fe(x))* and normal *(fn(x))* distribution for $\lambda = \mu = \alpha = 1$ and $a_1 = a_2 = 0.5$

where

$$\sum_{i=1}^{n} a_i = 1 \tag{1.74}$$

In (1.73) and (1.74) i indicates different distributions, a_i are the weighting factors and n is the number of combined distributions. The pdf of a combination of the exponential and normal distributions is displayed in Fig. 1.11.

Problems

1. The lifetime of a mechanical system is estimated to be Weibull distributed with $\alpha=10^4$ h, $\nu=0$ and $\beta=0.5$. If the mission time of the system is 500 h what is the probability that the mission will be successfully conducted given that the system was 300 h in service before the mission began? (Refer to (1.21) and (1.48)).

2. Determine the MTFF of the system from the preceding problem (Refer to (1.43) and (1.47)).

3. Calculate the pdf of the up time for the system in *Example 1.3* in terms of time and plot the associated graph. (Refer to (1.23)).

4. The residence time of a system in a certain state is uniformly distributed within interval (10 h, 20 h). Determine the probability that this state will be abandoned during time interval (10 h, 16 h). (Refer to (1.17) and (1.49)).

5. Consider a system consisting of three identical units. Initially, one unit is in service while the remaining two units are in cold standby mode. The unit in service is, when it fails, immediately replaced by a standby unit, if any is left. The up times of units are exponentially distributed with $\lambda=1$ fl./(10^4 h). What is the probability that the system will survive a year? (Refer to (1.68) bearing in mind that $\lambda=3\,\Theta$).

References

1. Gnedenko, B. V., *The Theory of Probability*, Mir Publishers, Moscow (1976)
2. Papoulis, A., *Probability, Random Variables and Stochastic Processes*, McGraw-Hill, London (1984)
3. Billinton, R., Allan, N. R., *Reliability Evaluation of Engineering Systems*, Plenum Press, New York (1992)
4. Shooman, M., *Probabilistic Reliability: An Engineering Approach*, McGraw-Hill, New York (1968)

5. Banks, J., Carson, J., *Discrete-Event System Simulation,* Prentice-Hall, New Jersey (1984)
6. Endrenyi, J., *Reliability Modeling in Electric Power Systems,* J. Wiley & Sons, New York (1978)
7. Ross M. S., *Introduction to Probability Models,* Academic Press, San Diego (1997)
8. Beichelt, F., Franken, P., *Zuverlässigkeit und Instandhaltung,* VEB Verlag Technik, Berlin (1983)
9. Janke-Emde-Lösch, *Tafeln Höherer Funktionen,* Teubner Verlag, Stuttgart (1960)
10. Barlow, R., Proschan, F., *Mathematical Theory of Reliability,* J. Wiley & Sons, New York (1965)

2 Renewable Two-State Systems

2.1
State-Transition Diagrams and Associated Indices

A system is considered renewable if, when it has failed, it is repaired as good as new or replaced by another system of the same type and performance. It is presumed that the repair or replacement time is a random variable with known Cdf. This Cdf may be deduced by observing the renewals of systems of the same type and by processing the acquired data on the renewal duration using the same approach as for the assessment of the system up time, as outlined in Chapter 1. We assume that the system has only two states: operating state and renewal state. Thus, the behavior of the system may be represented by a time- specific state-transition diagram as depicted in Fig. 2.1. Indices 1 and 2 represent the operating and the renewal states, respectively. Random variable X is the duration of sound system operation while random variable Y represents the renewal time. x_k and y_k are the operating and renewal times acquired during observation k. The aforesaid quantities are individual samples of the corresponding random variables X and Y.

The renewal time can be characterized by indices analogous to those for the operating time, as defined in Section 1.1. Thus, we define the *renewal rate* with regard to (1.13) as

$$\mu(y) = \frac{f(y)}{1 - F(y)} \tag{2.1}$$

with $F(y)$ and $f(y)$ being the Cdf and pdf of random variable Y. The quantity $\mu(y)dy$ is the probability that the renewal will be terminated during time interval $(y, y+dy)$ given that the system has not been renewed until y. The *mean renewal duration* equals, with regard to (1.16),

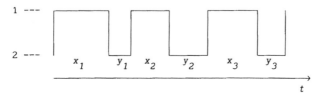

Fig. 2.1 Time-specific system state-transition diagram

$$r = E\{y\} = \int_0^\infty \Pr\{Y>y\}dy = \int_0^\infty [1-F(y)]dy \qquad (2.2)$$

Parameter r is commonly called the *Mean Time To Repair* (MTTR) [1,2].

The behavior of the system from the dependability point of view may be modeled by the state-transition diagram displayed in Fig. 2.2 with S and S' denoting the operating and under-renewal system state. It is implied that X and Y are generally distributed.

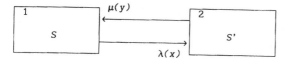

Fig. 2.2 System state-transition diagram

Consequently, the state-transition rates are presumed to depend upon the time the system has resided in a state before the transition to the other state. Generally, the transition rates may depend upon the absolute, chronological time too, if the system is presumed to be exposed to wearout with time. The deterioration of the system might increase the failure transition rate and decrease the repair rate. In such a case the transition rates should be treated as functions of both the state residence times and absolute time. However, the wearout effects are not pronounced during the technical life of a system, as discussed in Section 1.1, and may usually be ignored. Moreover, the aforementioned effects are not accounted for by the commonly accepted definition of the renewable systems given at the beginning of this chapter.

One of the most important dependability indices of engineering systems in practical applications is *system availability* $A(t)$ which is defined as the probability that the system will be sound at time instant t. The probability that the system will not be sound at time t, denoted as $U(t)$, is called *system unavailability*. The following relationships are simply deduced from the previous definitions

$$A(t) = p_1(t) = \Pr\{S\} \qquad (2.3)$$

$$U(t) = p_2(t) = \Pr\{S'\} \qquad (2.4)$$

$$A(t) + U(t) = 1 \qquad (2.5)$$

In (2.3) and (2.4) $p_k(t)$, $k=1,2$, denotes the probability that the system is in state k at time instant t..

It is clear that the probability that the system is sound at time t is higher for renewable than for nonrenewable systems. A nonrenewable system will be up at instant t only if it has not failed during the whole time period $(0,t)$. However, the renewable systems are renewed after each failure and might therefore be in the sound state after any number of faults if consecutive renewals have been carried out. From the aforesaid it follows that

$$A(t) > R(t) \tag{2.6}$$

$$U(t) < Q(t) \tag{2.7}$$

The availability and unavailability indices may generally be defined for non-renewable systems too. For these systems they coincide with the reliability and unreliability, respectively.

Availability and unavailability are dynamic parameters as they are time dependent. However, they differ from reliability and unreliability in that they have finite nontrivial steady-state values which are of great importance for the long-term system dependability evaluation.

2.2
Exponential Distribution of Up and Renewal Times

If both the sound and renewal state residence times are exponentially distributed, a relatively simple mathematical model can be composed for calculating the probabilities of system states.

The probability that the system will be in state 1 at time instant $t+dt$ equals, if dt is infinitely small

$$p_1(t+dt) = (1-\lambda dt)p_1(t) + \mu dt p_2(t) \tag{2.8}$$

The expression within the brackets in the first term on the r.h.s. of (2.8) yields the conditional probability that the failure will not occur during time interval dt given that the system has been sound until t, which follows from (1.40). As $p_1(t)$ is the probability that the system is sound at t, the first term in (2.8) is the probability that the system will stay in state 1 after dt. In the second term on the r.h.s. in (2.8) μdt is the conditional probability that the system will be renewed during time interval dt given that it was under renewal at time instant t and $p_2(t)$ is the probability that it was under renewal at t. Hence, this term is the probability that the system will be renewed during time interval dt.

By analogous reasoning as in composing relationship (2.8) one obtains

$$p_2(t+dt) = (1-\mu dt)p_2(t) + \lambda dt p_1(t) \tag{2.9}$$

The first term on the r.h.s. in (2.9) is the probability that the system will not be renewed during dt while the second term yields the probability that the system will fail during dt.

After elementary mathematical operations the following equations are derived from (2.8) and (2.9)

$$p_1'(t) = -\lambda p_1(t) + \mu p_2(t)$$
$$p_2'(t) = \lambda p_1(t) - \mu p_2(t)$$

$$(2.10)$$

As may be observed, the probabilities of the system being in sound and under-renewal states are calculable by solving the system of two linear differential equations with constant coefficients. Relationships (2.10) are *Kolmogorov equations* for the probabilities of states of the two-state systems under consideration [3]. As the Cdfs of the system state residence times are exponential and, consequently, the transition rates from system states do not depend on the state residence times, relationships (2.10) describe a *Markov stochastic process for discrete-state and time-continuous systems*. Since the transition rates do not depend on the absolute time, the process is *time-homogenous* [3,4].

By solving (2.10) we obtain

$$p_1(t) = \frac{\mu}{\lambda+\mu} + \frac{\lambda p_1(0)-\mu p_2(0)}{\lambda+\mu} \exp[-(\lambda+\mu)t]$$

$$(2.11)$$

$$p_2(t) = \frac{\lambda}{\lambda+\mu} + \frac{\mu p_2(0)-\lambda p_1(0)}{\lambda+\mu} \exp[-(\lambda+\mu)t]$$

$$(2.12)$$

where $p_1(0)$ and $p_2(0)$ are the initial values of the corresponding probabilities. Generally

$$p_1(t) + p_2(t) = 1$$

$$(2.13)$$

as states 1 and 2 are the only two states the system can encounter, by definition. Thus, (2.13) applies also to the initial probability values.

Assume that the system is up at $t=0$. Then

$$p_1(0) = 1, \qquad p_2(0) = 0.$$

$$(2.14)$$

For (2.14), expressions (2.11) and (2.12) become

$$p_1(t) = \frac{\mu}{\lambda+\mu} + \frac{\lambda}{\lambda+\mu} \exp[-(\lambda+\mu)t]$$

$$(2.15)$$

$$p_2(t) = \frac{\lambda}{\lambda+\mu} - \frac{\lambda}{\lambda+\mu} \exp[-(\lambda+\mu)t] \qquad (2.16)$$

The general pattern of probabilities $p_1(t)$ and $p_2(t)$ given by (2.15) and (2.16) is displayed in Fig. 2.3. The curves are plotted under the assumption that λ is much smaller than μ, which is typical for most engineering systems. As can be noted from (2.15) and (2.16) and from the diagrams in Fig. 2.2, the probabilities tend to certain limiting finite values when time t increases. These limiting values are the *steady-state probabilities*. They can formally be determined from (2.15) and (2.16) as

$$p_k = \lim_{t \to \infty} p_k(t) \qquad\qquad k=1,2 \qquad (2.17)$$

where p_1 and p_2 are the steady-state values of the system availability and unavailability which follow from (2.3) and (2.4)

$$A = p_1 , \qquad U = p_2 \qquad (2.18)$$

The steady-state availability can be interpreted as the probability that the system will be sound at an arbitrarily chosen time in the far future. As parameter μ is usually large, the steady-state values are reached in a relatively short time, after several weeks or, at most, several months.

From (2.15) and (2.16) we deduce, with regard to (2.17) and (2.18), that

$$A = \frac{\mu}{\lambda+\mu} \qquad (2.19)$$

$$U = \frac{\lambda}{\lambda+\mu} \qquad (2.20)$$

The steady-state values of system state probabilities can be directly determined from (2.10). As for steady-state solutions

$$\lim_{t \to \infty} p_k'(t) = 0 , \qquad\qquad k=1,2 \qquad (2.21)$$

equations (2.10) become, for steady state,

$$0 = -\lambda p_1 + \mu p_2$$
$$\qquad\qquad\qquad\qquad (2.22)$$
$$0 = \lambda p_1 - \mu p_2$$

As can be observed, the system of linear equations obtained is indeterminate as the equations are proportional to one another. To solve the system for p_1 and p_2 an equation should be replaced by the general relationship (2.13).

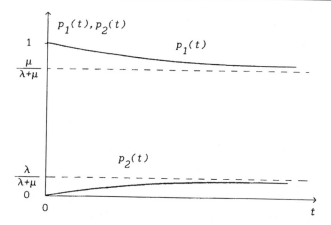

Fig. 2.3 Variation of system state probabilities with time

With reference to (1.13) the frequency of leaving a state is determinable as the product of the probability of residing in this state and the rate of transition out of it. Hence, for the case under consideration

$$f_1(t) = f_{12}(t) = \lambda p_1(t)$$
$$f_2(t) = f_{21}(t) = \mu p_2(t)$$

$$(2.23)$$

where index k, $k=1,2$, labels the frequency of abandoning state k and indices ik, $i,k=1,2$, $i \neq k$, label the frequency of transitions from state i to state k. As a two-state system is considered, the frequency of leaving a state equals the frequency of transition to the other state.

With reference to (2.23), expressions (2.10) can be written in the form

$$p_1'(t) = -f_1(t) + f_{21}(t)$$
$$p_2'(t) = -f_2(t) + f_{12}(t)$$

$$(2.24)$$

For steady state (2.24) yields

$$- f_1 + f_{21} = 0$$
$$- f_2 + f_{21} = 0$$

$$(2.25)$$

As may be noted from (2.25), in steady-state conditions there is a frequency balance for each system state. That is, the frequency of leaving a state is equal to the frequency of encountering this state. As seen clearly from (2.24), there is no such balance in the transient phase. If we insert p_1 and p_2 into (2.23) according to (2.19) and (2.20) the following expressions are obtainable

$$f_1 = f_2 = \frac{\lambda\mu}{\lambda+\mu} \tag{2.26}$$

Example 2.1

Consider a system having constant failure transition rate $\lambda=1$ fl./(10^3 h) and renewal transition rate $\mu=0.02$ ren./h. The task is to determine the characteristic reliability indices of the system as a function of time assuming that the system is sound at $t=0$.

As transition rates are constants, both the sound and renewal states are exponentially distributed. Thus, we can use the expressions derived in this section. The results of the calculations are presented in Table 2.1. As can be seen from Table 2.1, all indices analyzed reach their steady-state values after approximately 400 h. After the aforementioned time period has expired, the frequencies $f_1(t)$ and $f_2(t)$ become practically equal to $f_{21}(t)$ and $f_{12}(t)$, satisfying the frequency balance relationships.

Table 2.1. Calculated dependability indices

t h	$A(t)=p_1(t)$	$U(t)=p_2(t)$	$f_2(t)=f_{12}(t)$ fl./(10^4 h)	$f_1(t)=f_{21}(t)$ ren./(10^4 h)
0	1.0000	0.0000	10.00	0.00
10	0.9910	0.0090	9.91	1.80
30	0.9777	0.0223	9.77	4.45
50	0.9708	0.0292	9.71	5.84
100	0.9691	0.0309	9.69	6.18
200	0.9531	0.0469	9.53	9.38
300	0.9525	0.0475	9.52	9.50
400	0.9524	0.0476	9.52	9.52

□

2.3
Absorbing States and the Equivalent Steady-State Transition Diagram

The Cdf of the residence time in a given system state (set of system states) can be deduced from the system state-transition diagram by modeling all the remaining system states as *absorbing states*. Absorbing states are those from which no transitions to other states can occur. Thus, any system state can be converted into an absorbing state by canceling all transitions from this state to another states. If all system states excluding the state under consideration are absorbing, then the system equations describe the process of abandoning the system state under consideration and the absolute, chronological time is the residence time in this state. The probability that the system is in an absorbing state at a time t is the probability that the system has abandoned the state under consideration before or at t. Hence, by definition, the variation of the aforementioned probability with time coincides with the Cdf of the residence time in the state under consideration.

The previously described approach for analyzing the state residence times will be applied to the two-state system investigated in this chapter, for illustration. It is clear that this approach may also be applied to systems with multiple states and with general residence time distributions.

In order to determine the Cdf of the sound system state of the system under consideration, state 2 is converted into an absorbing state. The system state-transition diagram for this case is depicted in Fig. 2.4.

Fig. 2.4 System state-transition diagram with the renewal state being absorbing state

The relationship associated with the probability of state 1 is

$$p_1'(t) = -\lambda p_1(t) \tag{2.27}$$

For $p_1(0)=1$, equation (2.27) yields

$$p_1(t) = \exp(-\lambda t) \tag{2.28}$$

If the duration of sound operation is denoted by X then, obviously,

$$\Pr\{X>t\} = p_1(t)$$

$$\Pr\{X\leq t\} = 1-p_1(t) = p_2(t) \tag{2.29}$$

The mean time to failure equals (Section 1.1 and (2.28) and (2.29))

$$\text{MTTF} = \int_0^\infty p_1(t)dt = \frac{1}{\lambda} = m \qquad (2.30)$$

As it is implied that the system is as good as new after each renewal, λ does not change with renewals and the MTTF coincides with the MTFF.

In order to deduce the Cdf of the renewal time, state 1 is converted into an absorbing state, as shown in Fig. 2.5.

The expression for the probability of state 2, as obtained from Fig. 2.2, is

$$p_2'(t) = - \mu p_2(t) \qquad (2.31)$$

Fig. 2.5 State-transition diagram with state 1 being absorbing state

By solving (2.31) for $p_2(t)$ we obtain, under the assumption $p_2(0)=1$,

$$p_2(t) = \exp(-\mu t) \qquad (2.32)$$

It is clear that $p_2(t)$ coincides with probability $\Pr\{Y>t\}$ where Y is the duration of residence in state 2. From (2.32) and (2.2) it follows that

$$\text{MTTR} = r = \frac{1}{\mu} \qquad (2.33)$$

The steady-state behavior of the system may be modeled by the equivalent cyclic symmetrical diagram displayed in Fig. 2.6. This statement can be simply proved. The probability that the system represented by the diagram in Fig. 2.6 will be up at a randomly chosen observation instant in the far future, which is the definition of the steady-state availability value, equals

$$A = \frac{m}{m+r} \qquad (2.34)$$

For the probability that the system is in the renewal state, which is by definition the steady-state unavailability, we obtain from Fig. 2.6

$$U = \frac{r}{m+r} \qquad (2.35)$$

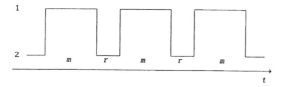

Fig. 2.6 Equivalent state-transition diagram

If we replace m and r in (2.34) and (2.35) with regard to (2.30) and (2.33), the expressions (2.19) and (2.20) are obtained, which proves our statement.

With regard to Fig. 2.6 the *mean time between failures* (MTBF) equals

$$T = m + r \tag{2.36}$$

The failure frequency (frequency of leaving state 1) equals the inverse of T

$$f_1 = \frac{1}{T} = \frac{1}{m+r} \tag{2.37}$$

This frequency coincides with the frequency of the transitions from renewal to operating state f_{21}. If we substitute in (2.37) r and m according to (2.19) and (2.20) relationships (2.26) are obtained.

Let us assume that n operation–renewal cycles happened during a period Θ of observation of system steady-state behavior. The total expected renewal duration during this period equals

$$\tau = n\,r \tag{2.38}$$

With regard to (2.35) we can write

$$U = \frac{n\,r}{n\,(m+R)} = \frac{\tau}{\Theta} \tag{2.39}$$

Expression (2.39) shows that the steady-state value of the unavailability provides a simple way to calculate the expected cumulative duration of the renewal of a system during a time period of interest given that it can be taken that the system was in its steady state at the beginning of this period. In (2.38) and (2.39) n may be any number, not necessarily an integer. It means that time period Θ may be selected independently of the number of state cycles and their duration.

Example 2.2
We shall determine the steady-state values of all the main dependability indices for the system in *Example 2.1* including the expected cumulative renewal duration during a year.

In accordance with (2.30) and (2.33) we have

$$m = \frac{1}{10^{-3}} = 1000 \text{ h} , \quad r = \frac{1}{0.02} = 50 \text{ h}$$

Eqs. (2.34) and (2.35) yield

$$A = \frac{1000}{1000+50} = 0.9524 , \quad U = \frac{50}{1000+50} = 0.0476$$

From (2.36) and (2.37) it follows

$$T = 1000 + 50 = 1050 \text{ h} , \quad f_1 = \frac{1}{1050} = 0.9524 \text{ fl/(1000 h)}$$

$$\tau = 0.0476 \times 8760 = 416.98 \text{ h}$$

$\square$

2.4
General Cdf of Renewal Duration

The exponential Cdf of renewal duration implies that the probability of renewal during time interval $(y, y+dy)$ does not depend on time y already expended for renewal, which is not realistic even for a quite inexperienced repair crew. For this reason, in this section we shall consider some methods enabling the evaluation of transient and steady-state dependability indices of systems with generally distributed renewal time. The state-transition diagram of such a system is depicted in Fig. 2.7.

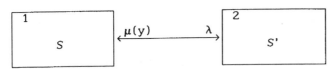

Fig. 2.7 State-transition diagram for a system with generally distributed renewal time

2.4.1
Method of Supplementary Variables

The expressions for the probabilities of residing in operating and renewal states are complex because the frequency of the transition from state 2 to state 1 depends on the distribution of the time spent in state 2. This is clarified in Fig. 2.7 by indicating that the transition rate to state 1 depends on y. The probability of being in a renewal state at a time t depends in a complex way on both the operation and renewal processes as it is affected by the probability that the renewal state is encountered at an instant before t. However, the mathematical model of the system under consideration can be built in a relatively simply form if the time spent in a renewal state is introduced as a supplementary, separate random variable [5]. Then the time-specific frequency of transitions from state 2 to state 1 is

$$f_{21}(t) = \int_0^t \mu(y)\pi_2(t,y)dy \qquad (2.40)$$

where $\pi_2(t,y)dy$ is the probability that the system is in state 2 at t and that it has been in that state for time y. It is implied that $p_1(0)=1$.

On the basis of (2.24) and (2.40) we can write

$$p_1'(t) = -\lambda p_1(t) + \int_0^t \mu(y)\pi_2(t,y)dy \qquad (2.41)$$

The increment of $\pi_2(t,y)$ during a small time increment dt equals

$$\frac{\partial \pi_2(t,y)}{\partial t} dt + \frac{\partial \pi_2(t,y)}{\partial y} dt = -\mu(y)\pi_2(t,y)dt \qquad (2.42)$$

The term on the l.h.s. of (2.42) is the total differential of a function of two variables. By definition, y is the time spent in state 2 measured from the instant of the arrival in this state while t is the time counted from the beginning of the observation. If the system is in state 2 at t then a small increment of t will cause the same increment of y. The term on the r.h.s. of (2.42) takes into account that during time increment dt function $\pi_2(t,y)$ will decrease due to renewal activity. If (2.42) is divided by dt a partial differential equation is obtained. It is important to note that

$$\pi_2(t,0) = \lambda p_1(t) \qquad (2.43)$$

as the frequency of encountering state 2 equals to the frequency of leaving state 1.

Also, by definition of $\pi_2(t,y)$

$$p_2(t) = \int_0^t \pi_2(t,y)dy \qquad (2.44)$$

Relationships (2.41) to (2.44) build a closed system for determining $p_1(t)$ and $p_2(t)$ which can be solved by applying various approaches [4–7].

The steady-state solutions are obtainable from the previously derived expressions as limiting values for t tending to infinity. The steady-state values will again be denoted using the same symbols as for the time-specific values but by omitting argument t.

For the steady state, expressions (2.41) to (2.44) become

$$0 = -\lambda p_1 + \int_0^\infty \mu(y)\pi_2(y)dy \tag{2.45}$$

$$\frac{\partial \pi_2(y)}{\partial y} = -\mu(y)\pi_2(y) \tag{2.46}$$

$$\pi_2(0) = \lambda p_1 \tag{2.47}$$

$$p_2 = \int_0^\infty \pi_2(y)dy \tag{2.48}$$

The system of equations should be completed by the general relationship

$$p_1 + p_2 = 1 \tag{2.49}$$

By solving (2.46) we obtain

$$\pi_2(y) = \pi_2(0) \exp\left(-\int_0^y \mu(u)du\right) \tag{2.50}$$

If $\pi_2(y)$ from (2.50) is inserted into (2.48) the following expression is obtained, bearing in mind (2.47),

$$p_2 = \lambda p_1 r \tag{2.51}$$

where

$$r = \int_0^\infty \{\exp[-\int_0^y \mu(u)du]\} \, dy \tag{2.52}$$

Analogous to (1.14) the integrand in (2.52) is the probability that the system will reside in state 2 until y. Consequently, taking into account (2.2), parameter r is the mean renewal duration. If the equivalent renewal rate is introduced

$$\mu = \frac{1}{r} \tag{2.53}$$

expression (2.51) converts to the same form as for an exponential renewal time distribution. The same is also the case with the expressions for the other dependability indices. From the aforesaid we conclude that the steady-state dependability indices could be calculated from the exponential model and the associated Kolomogorov equations if μ is used according to (2.52) and (2.53). However, such a conclusion is not valid for all engineering systems having nonexponential state residence time distributions [8,9]. The latter holds generally if the activity with nonexponential distribution is initialized and terminated at the same system state, as it was in the case analyzed.

Example 2.3

Consider a system having a Weibull-distributed renewal time with $\alpha = 5$ h, $\nu = 0$ and $\beta = 2$. The system up state duration is exponentially distributed with $\lambda = 1$ fl. $/(1000$ h). The main steady-state dependability indices have to be determined for this system.

The mean duration of renewal equals, with regard to (1.43),

$$r = \alpha\Gamma(1+\frac{1}{\beta}) = 5\times0.8862 = 4.4301 \text{ h}$$

With regard to (2.53) we have

$$\mu = \frac{1}{4.4301} = 0.2257 \text{ ren./h}$$

and then

$$P_1 = A = \frac{\mu}{\lambda+\mu} = \frac{0.2257}{10^{-3}+ 0.2257} = 0.9956$$

$$P_2 = U = 1 - A = 4.411\times 10^{-3}$$

$$\text{MTBF} = \frac{1}{\lambda} + r = 1004.43 \text{ h}$$

□

2.4.2
Fictitious States

Many renewal time distributions encountered in practice may be fairly well modeled by a combination of states having exponentially distributed residence times. Such states are fictitious as they are individually not related to any actual system state.

Their combination and parameters are selected to generate the Cdf of the entire time spent in this combination to be as close as possible to the Cdf of the real renewal state. By introducing the fictitious states the system is converted into a Markov system and, thus, can be described using Kolomogorov equations. These equations can be then handled by applying well developed and comparatively simple mathematical methods, as will be discussed in Chapter 3. The number of fictitious states should be kept as low as possible in order to limit the state space of the system under consideration.

The basic criterion used to determine the parameters of a selected combination of states might be

$$M_i = M_{ei}, \qquad i=1,...,m \qquad (2.54)$$

with M_i and M_{ei} being ith moments of the real and equivalent renewal time distributions. By definition, the ith moment of a probability distribution is [10]

$$M_i = \int_0^\infty y^i f(y)dy \qquad (2.55)$$

In (2.55) y takes the possible values of the random variable and $f(y)$ is its pdf.

As follows from (2.55), the first moment of random variable y is its mathematical expectation, i.e. the mean value from the probability point of view.

General expression (2.55) applies to fictitious states also

$$M_{ei} = \int_0^\infty y^i f_e(y)dy \qquad (2.56)$$

In (2.56), $f_e(y)$ is now the pdf of the entire time of residence in the set of fictitious states.

Parameter m in (2.54) is usually selected to be equal to the number of the unknown parameters of the chosen combination of fictitious states. Then (2.54) provides a sufficient number of equations to determine all these parameters. However, if we adopt some parameters in advance, number m is adequately reduced.

The expressions for M_{ei} can be readily obtained by applying the *Laplace transform* (*L*-transform).

By definition, the *L*-transform of $f_e(y)$ is [11]

$$f_e(s) = \int_0^\infty \exp(-sy)f_e(y)dy \qquad (2.57)$$

with $f_e(s)$ being the *L*-transform of $f_e(y)$ and s denoting the operator of this transform.

The ith derivative of (2.57) with regard to s equals

$$f_e(s)^{(i)} = (-1)^i \int_0^\infty \exp(-sy)\ y^i\ f_e(y)dy \tag{2.58}$$

From (2.56) and (2.58) it follows that

$$M_{ei} = (-1)^i\ (f_e(0))^{(i)} \tag{2.59}$$

where $f_e(0)^{(i)}$ denotes the ith derivative of $f_e(s)$ with regard to s, for $s=0$.

Pdf $f_e(y)$ is determined by converting the system operating state into an absorbing state implying that the renewal state is encountered at $t=0$. Then, as discussed in Section 2.3, the absolute time t coincides with time y of residence in the fictitious states. Consequently

$$\sum_k p_k(y) = \Pr\{Y > y\} = 1 - \Pr\{Y \le y\} \tag{2.60}$$

with Y being the total residence time in fictitious states and $p_k(y)$ denoting the probabilities of residing in fictitious states. Index k applies to all fictitious states.

With regard to (1.29) and (2.60) we can write

$$f_e(y) = -\sum_k p_k'(y) \tag{2.61}$$

By applying the L-transform to (2.61) the following expression is derived

$$f_e(s) = \sum_k p_k(0) - s\sum_k p_k(s) = 1 - s\sum_k p_k(s) \tag{2.62}$$

where $p_k(0)$ is the probability of being in fictitious state k at time 0, that is at the instant of encountering the set of fictitious states. Hence, the sum of these probabilities for fictitious states equals 1. The expressions for the L-transforms $p_k(s)$ are obtainable from the corresponding Kolmogorov's equations describing the combination of the fictitious states as will be demonstrated for two characteristic cases later in this section.

As follows from (2.62) and (2.59), the mean duration of residence in the set of fictitious states is

$$M_{e1} = \sum_k p_k(s) \quad \text{for } s = 0 \tag{2.63}$$

The application of the method with fictitious states will be demonstrated for the two-state system considered in this chapter. The fictitious states will be used to model the renewal state.

Example 2.4
The renewal time of a system is uniformly distributed within interval (8 h,12 h). The first three moments of this distribution are to be determined.

With regard to (1.50) and (2.55) the ith moment of the uniformly distributed renewal time equals

$$M_i = \int_a^b \frac{y^i dy}{b-a} = \frac{1}{i+1} \frac{b^{i+1} - a^{i+1}}{b-a}$$

Expression (1.51) is obtainable from the derived general formula as a special case, for $i=1$.

After inserting the values of a and b the following results are obtained

$$M_1 = 10.0 \text{ h}, \quad M_2 = 101.3 \text{ h}^2, \quad M_3 = 1040.0 \text{ h}^3$$

□

2.4.3
Series Connection of Fictitious States

Suppose that the renewal state is modeled by a series connection of n fictitious states as shown in Fig. 2.8. The fictitious states are labeled as states 2.1,...,2.n. Thus, the actual renewal residence time will be approximated by the sum of n exponentially distributed residence times of fictitious states. The relevant Cdf and pdf of the total residence time in fictitious states are obtainable by presuming that the up system state is absorbing and that the renewal state, i.e. state 2.1 of the series connection, is encountered at the initial time instant.

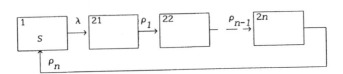

Fig. 2.8 System state-transition diagram with renewal state approximated by a series connection of fictitious states

For the case considered Kolomogorov equations for fictitious states are

$$p'_{2.1}(y) = -\rho_1 p_{2.1}(y) \tag{2.64}$$

$$p'_{2.k+1}(y) = -\rho_{k+1} \, p_{2.k+1}(y) + \rho_k \, p_{2.k}(y), \quad k=2,...,n-1 \tag{2.65}$$

The application of the L-transform to (2.64) and (2.65) yields

$$p_{2.1}(s) = \frac{1}{s+\rho_1} \tag{2.66}$$

$$p_{2.k+1}(s) = \frac{\rho_k}{s+\rho_{k+1}} \, p_{2.k}(s), \quad k=1,...,n-1 \tag{2.67}$$

With reference to (2.66) and to the recurrent expression (2.67) the following expressions for the L-transforms of the probabilities of residing in fictitious states are deduced

$$p_{2.k}(s) = \prod_{j=1}^{k} \frac{\rho_{j-1}}{s+\rho_j}, \quad \rho_0 = 1 \tag{2.68}$$

The probability of residing in a fictitious state at time instant y is obtainable by conversion of (2.68) into the time domain.

If the L-transform of a function is the quotient of two polynomials

$$F(s) = \frac{A(s)}{B(s)} \tag{2.69}$$

then its conversion into the time domain is [13]

$$F(y) = \sum_i \frac{1}{(n_i-1)!} \lim_{s \to a_i} \left((s-a_i)^{n_i} \frac{A(s)}{B(s)} \exp(sy) \right)^{(n_i-1)} \tag{2.70}$$

a_i is root i of $B(s)$ while n_i is the multiplicity of this root. Index i is over all distinct roots. Superscript (n_i-1) denotes the order of the derivative with regard to s.

Let us introduce the polynomial

$$B_i(s) = \frac{B(s)}{(s-a_i)^{n_i}} \tag{2.71}$$

Then, (2.70) can be written in the form

$$
F(y) = \sum_i \frac{1}{(n_i-1)!} \left(\frac{A(s)}{B_i(s)} \exp(sy) \right)_{s=a_i}^{(n_i-1)}
\tag{2.72}
$$

The subscript to the r.h.s. bracket indicates that the derivative value for $s=a_i$ should be inserted.

If we apply (2.72) to (2.68), under the assumption $n_i=1$ for all i, the following expressions are obtained

$$
P_{2,k}(y) = \sum_{i=1}^{k} \frac{\exp(-\rho_i y)}{B_i(-\rho_i)} \prod_{j=1}^{k} \rho_{j-1}
$$

$$
B_i(-\rho_i) = \prod_{j=1}^{k} (\rho_j - \rho_i) \quad j \neq i
\tag{2.73}
$$

The Cdf of the residence time in fictitious states is

$$
F(y) = \Pr\{Y \le y\} = 1 - \sum_{k=1}^{n} P_{2,k}(y)
\tag{2.74}
$$

As follows from (2.63), the mean renewal duration generated by the series connection equals

$$
r_e = M_{el} = \sum_{k=1}^{n} P_{2,k}(s) \quad \text{for } s = 0
\tag{2.75}
$$

With reference to (2.68) we have

$$
P_{2,k}(0) = \prod_{j=1}^{k} \frac{\rho_{j-1}}{\rho_j} = \frac{1}{\rho_k}
\tag{2.76}
$$

Consequently

$$
r_e = \sum_{k=1}^{n} \frac{1}{\rho_k}
\tag{2.77}
$$

Example 2.5

Let us determine moments M_{e1} and M_{e2} for two series-connected states approximating the system renewal state.

With reference to (2.77) we have

$$M_{e1} = \frac{1}{\rho_1} + \frac{1}{\rho_2}$$

From (2.62) it follows that

$$f_e^{(1)}(s) = -\sum_{k=1}^{2} P_{2k}(s) - s\sum_{k=1}^{2} p_{2k}^{(1)}(s)$$

$$f_e^{(2)}(s) = -2\sum_{k=1}^{2} p_{2k}^{(1)}(s) - s\sum_{k=1}^{2} p_{2k}^{(2)}(s)$$

Then, according to (2.59)

$$M_{e2} = -2\sum_{k=1}^{2} p_{2k}^{(1)}(s) \quad \text{for } s=0$$

Taking account of (2.66) we have

$$p_{2.1}^{(1)}(s) = -\frac{1}{(s+\rho_1)^2} \,, \qquad p_{2.1}^{(1)}(0) = -\frac{1}{\rho_1^2}$$

From (2.68) it follows that

$$p_{2.2}^{(1)}(s) = \left(\frac{1}{s+\rho_1}\frac{\rho_1}{s+\rho_2}\right)^{(1)} = -\frac{2s\rho_1 + (\rho_1+\rho_2)\rho_1}{(s+\rho_1)^2\,(s+\rho_2)^2}$$

$$p_{2.2}^{(1)}(0) = -\left(\frac{1}{\rho_2^2} + \frac{1}{\rho_1\rho_2}\right)$$

After inserting the expressions derived for $p^{(1)}_{2.1}$ and $p^{(1)}_{2.2}$ in the expression for M_{e2} we obtain

$$M_{e2} = 2\left(\frac{1}{\rho_1^2} + \frac{1}{\rho_1\rho_2} + \frac{1}{\rho_2^2}\right)$$

□

2.4.4
Parallel Connection of Fictitious States

The renewal state can be approximated by a set of n mutually independent parallel
fictitious states as shown in Fig. 2.9. The probability that, given that the system has
failed, it will transit to fictitious state 2.i is denoted by ω_i where

$$\sum_{i=1}^{n} \omega_i = 1 \qquad (2.78)$$

Parameter λ is the system failure transition rate.

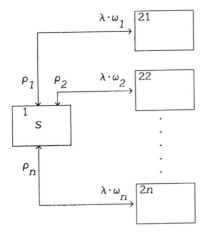

Fig. 2.9 Parallel combination of fictitious states

If we convert state 1 into an absorbing state, Kolomogorov equations for fictitious
states become

$$p_{2.k}'(y) = -\rho_k\, p_{2.k}(y), \qquad k=1,...,n \qquad (2.79)$$

By applying the L-transform to (2.79) we obtain

$$p_{2.k}(s) = \frac{\omega_k}{s+\rho_k} \qquad (2.80)$$

bearing in mind that ω_k is the probability of encountering state 2.k after leaving
state 1. The conversion of (2.80) into the time domain yields

$$p_{2.k}(t) = \omega_k \exp(-\rho_k y) \qquad (2.81)$$

Consequently, the Cdf of the residence time in fictitious states is

$$F(y) = \Pr\{Y \le y\} = 1 - \sum_{k=1}^{n} \omega_k \exp(-\rho_k y) \qquad (2.82)$$

and the corresponding pdf is

$$f(y) = \sum_{k=1}^{n} \rho_k \omega_k \exp(-\rho_k y) \qquad (2.83)$$

From (2.80) and (2.63) we deduce that the equivalent MTTR generated by the fictitious states equals

$$r_e = \sum_{k=1}^{n} \frac{\omega_k}{\rho_k} \qquad (2.84)$$

As can be observed, the parallel combination is associated with n independent transition rates and $n-1$ independent initial probabilities ω_i. That provides a good possibility to approximate various renewal time distributions. However, the pdf of the parallel combination has a nonzero initial value, as is obvious from (2.83), which limits its application to distributions having the same property.

It is clear that various other combinations of fictitious states may be designed for a proper approximation of nonexponentially distributed renewal times [9].

Example 2.6
Consider a system having both the up and renewal state residence times nonexponentially distributed. The up state should be modeled by two series connected states with transition rates ρ_{11} and ρ_{12}. The renewal time distribution should be approximated by two parallel fictitious states having initial probabilities ω_{21} and ω_{22} which transit to the up state with rates ρ_{21} and ρ_{22}.

The state transition diagram for the case under consideration is depicted in Fig. 2.10. Numerals 1.1. and 1.2 denote the fictitious states modeling the up state

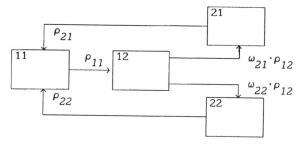

Fig. 2.10 System model with two series and two parallel fictitious states

while numerals 2.1 and 2.2 are related to the fictitious renewal states. As observed, the transition occurs from state 1.1 to state 1.2 and from this state to states 2.1 and 2.2 in accordance with the corresponding initial probabilities of these states. Renewal states transit to state 1.1 with corresponding transition rates.

Example 2.7
The up state of the system in *Example 2.6* should be modeled by two parallel fictitious states having failure transition rates ρ_{11} and ρ_{12}. The probabilities of encountering the fictitious up states from states 2.1 and 2.2, after the renewal is terminated, are ω_{11} and ω_{12}. The state transition diagram for this case is displayed in Fig. 2.11.

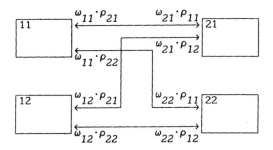

Fig. 2.11 System model with two pairs of parallel fictitious states

□

2.4.5
Open Loop Sequence of Consecutive States

If a short-term dependability evaluation of a system is of interest, the approach based upon the open loop system representation displayed in Fig. 2.12 may be used. The system is modeled as a series of consecutive alternating up and renewal states. States 1.1, 1.2, ..., are the initial up state and, consecutively, up states after first renewal, second renewal, etc. States 2.1, 2.2,... are the renewal states after the first failure, second failure, etc. The probabilities of consecutive states can be calculated by applying the recurrent relationships.

The probability of being in state 1.k at absolute time instant t equals

$$p_{1.k}(t) = \int_0^t S_1(t-x)\, p_{2.k-1}(x)\, \mu(x)dx \tag{2.85}$$

where $S_1(z)$ designates the probability of staying in up state for time z. By definition, this function, usually called *the survival function*, is the complement of the Cdf of

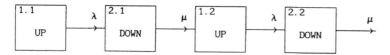

Fig. 2.12 Open loop state-transition diagram

system up time. Term $p_{2.k-1}(x)\mu(x)dx$ gives the probability that the system is renewed and restored to the up state at time x while term $S_1(t-x)$ is the probability that the system will stay in the up state until t.

The probability of being in state 2.k at time instant t can be calculated as

$$p_{2.k}(t) = \int_0^t S_2(t-x)\, p_{1.k}(x)\lambda(x)dx \qquad (2.86)$$

with the meaning of the associated terms analogous to those in (2.85). $S_2(z)$ is the probability of staying in the renewal state for time z. This function is the complement of the Cdf of the renewal time.

As may be seen, the previously derived recurrent relationships are valid and applicable for any Cdf of up and renewal times. Moreover, they allow for changing the transition rates with the number of transitions, if needed to account for deterioration. It is important to note that $p_{1.k}(t)$ is the probability that the system failed and was renewed $k-1$ times until t, under the assumption that it was up at $t=0$, which is useful information.

The probability that the system will be sound at time t is obtainable by summing probabilities $p_{1.k}(t)$ with k tending to infinity. However, in practical applications the number of states should be limited to a certain number, say m, providing the required accuracy of calculation. The bounds of $p_1(t)$ are

$$\sum_1^m p_{1.k}(t) < p_1(t) < \sum_1^m p_{1.k}(t) + p_{1.m+1}^*(t) \qquad (2.87)$$

with $p_{1.m+1}^*(t)$ being the probability of up state 1.$m+1$ when considered as an absorbing state

$$p_{1.m+1}^*(t) = \int_0^t p_{2.m}(x)\mu(x)dx \qquad (2.88)$$

Example 2.8
Consider a system having exponentially distributed up and renewal states with associated transition rates λ and μ. This system can be analyzed using

Kolomogorov's equations as shown in Section 2.2. The open loop approach will be applied here, for demonstration and comparison. It is implied that the system is sound at $t=0$.

The probability of residing in the up state before the first failure equals

$$p_{1.1}(t) = \exp(-\lambda t) \qquad (2.89)$$

The probability of being under renewal after the first failure is, with reference to (2.86)

$$
\begin{aligned}
p_{2.1}(t) &= \int_0^t \exp(-\mu(t-x))\exp(-\lambda x)\lambda dx \\
&= \frac{\lambda}{\mu-\lambda} \left[\exp(-\lambda t) - \exp(-\mu t) \right]
\end{aligned} \qquad (2.90)
$$

The probability of being in the up state after the first renewal is

$$
\begin{aligned}
p_{1.2}(t) &= \frac{\lambda}{\mu-\lambda} \int_0^t \exp(-\lambda(t-x))[\exp(-\lambda x)-\exp(-\mu x)]\mu dx \\
&= \frac{\lambda\mu}{\mu-\lambda}t \, \exp(-\lambda t) + \frac{\lambda\mu}{(\mu-\lambda)^2}[\exp(-\lambda t)-\exp(-\mu t)]
\end{aligned} \qquad (2.91)
$$

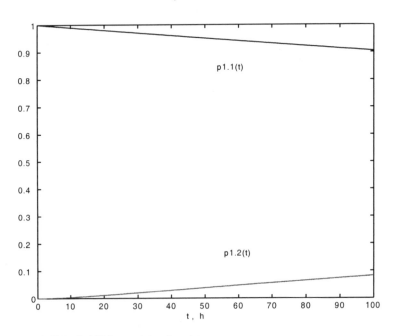

Fig. 2.13 Probabilities $p_{1.1}(t)$ and $p_{1.2}(t)$

Fig. 2.13 displays the results of the open loop calculation for a system with $\lambda=1$ fl./1000 h and $\mu=1$ ren./10 h. In Fig. 2.14 these results are compared with the solution obtained using exact expressions (2.15) and (2.16). As may be seen from Fig. 2.14, the approximate solution is very close to the exact one for moderate t. For $t=50$ h the error is 0.05%. For longer times the error increases and a greater number of consecutive up states should be considered.

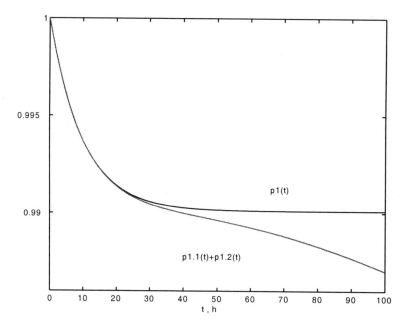

Fig. 2.14 Exact and approximate probability of the up system state

□

Problems

1. The renewal duration of a system is lognormally distributed with $\mu=\ln2$ and $\alpha=1$ if random variable X is measured in hours. Determine the MTTR and the steady-state availability of the system if the failure rate is $\lambda=1$ fl./500 h. (Refer to (1.71) and *Example 2.3*)

2. By applying (2.59) check the results obtained in *Example 2.4*.

3. Approximate the uniform renewal time distribution in *Example 2.4* by two series-connected fictitious states designed to satisfy the condition $M_{ei}=M_i$ $i=1,2$. (Use the expressions derived in *Example 2.5*)

4. Determine the second moment of the renewal time distribution approximated by two parallel connected fictitious states.

5. Using (2.87) assess the error in approximating by $p_{1.1}(t)$ the probability of the up state of the system in *Example 2.8*.

References

1. Henley, E.J., Kumamoto, H., *Reliability Engineering and Risk Assessment*, Prentice-Hall, Englewood Cliffs (1981)
2. Billinton, R., Allan R.N., *Reliability Evaluation of Engineering Systems*, Plenum Press, New York (1992)
3. *Problems of the Mathematical Theory of Dependability*, (in Russian) Edited by Gnedenko,V.B., Radio i Sviaz, Moscow (1983)
4. Cox, D.,R., Miller, H.,D., *The Theory of Stochastic Processes*, Chapman and Hall, London (1965)
5. Cox, D., R. , The analysis of non-Markovian stohastic processes by the inclusion of supplementary variables, *Proc. Cambridge Philos. Soc.*, **51** (1955), pp. 433–441.
6. Matveiev, V.F., Ushakov, V.G., *Mass Servicing Systems,* (in Russian) Moscow University Press, Moscow (1984)
7. Saaty, T.L., *Elements of Queuing Theory,* McGraw-Hill, New York (1961)
8. Singh, C., Billinton, R., Reliability modelling in systems with non-exponential down time distributions, *IEEE Trans.* **PAS-92** (1973), pp.790–800.
9. Singh, C., Billinton, R. *System Reliability Modelling and Evaluation*, Hutchinson & Company, London (1977)
10. Feller, W. *An Introduction to Probability Theory and Its Applications*, Vols. **I,II**, J. Wiley & Sons, New York (1968,1971)
11. Doetsch, G., *Anleitung zum praktischen Gebrauch der Laplace-Transformation und der Z-Transformation*, R.Oldenbourg, München (1967)
12. Cox, D., *Renewal Theory*, Methuen & Company, London (1962)
13. Nahman, J., *Methods for the Dependability Analysis of Electric Power Systems* (in Serbian), Naucna Knjiga, Belgrade (1992)

3 Markov Systems

Let us consider systems having an arbitrary but finite number of distinct states with exponentially distributed residence times. Suppose also that the interstate transition rates are independent of the absolute time. Such systems are usually called *dicrete-state time-homogenous and continuous Markov systems*. The two-state system discussed in the previous chapter is an elementary example of such systems. The dependability of many technical apparatuses and processes may be successfully evaluated using the Markov model. Systems having some non-exponentially distributed state residence times may also be analyzed if converted to an equivalent Markov system by replacing the nonexponential states by appropriate sets of fictitious states as described in Section 2.4.

3.1
Equations Determining the Probability of System States

With regard to the properties of the systems under consideration, the probabilities of system states are obtainable by solving the corresponding system of Kolmogorov equations

$$p_k'(t) = \sum_{i=1}^{k-1} a_{ki} p_i(t) + a_{kk} p_k(t) + \sum_{i=k+1}^{n} a_{ki} p_i(t) \quad \forall k \tag{3.1}$$

For $k=1$ the first term on the l.h.s. in (3.1) is zero.

Symbols in (3.1) have the following meaning:

$p_k(t)$ - probability of the system being in state k at time instant t

n - total number of system states

a_{ki} - transition rate from state i to state k, $k \neq i$

a_{kk} - negative value of the transition rate from state k to another states

The transition rate of abandoning state k equals the sum of the rates of transition to the remaining system states

$$a_{kk} = -\sum_{j=1}^{n} a_{jk} \quad j \neq k \tag{3.2}$$

If the system under consideration is built from m components which individually might encounter a number of various states, the following holds for the total number of system states

$$n \leq \prod_{j=1}^{m} n_j \tag{3.3}$$

with n_j designating the number of states of component j. It is clear that the term on the r.h.s. of (3.3) is the total number of the combinations of components states. However, it may happen that some of those combinations are not possible for various reasons, as will be discussed later in this chapter. This explains the "less than or equal to" sign in (3.3).

Equations (3.1) can be written in matrix form

$$[p'(t)] = [a][p(t)] \tag{3.4}$$

where

$$[p(t)] = [p_1(t)...p_n(t)]^T \tag{3.5}$$

$$[a] = [a_{ik}] \qquad i,k=1,...,n \tag{3.6}$$

The state probabilities are obtainable by solving the system of differential equations (3.4) for known initial probability values $p_k(0)$, $k=1,...,n$. These equations can be solved by applying any of the well developed numerical methods for solving ordinary differential equations, such as the Runge–Kutta method [1], for example.

An explicit form of solutions can be obtained by applying the L-transform to (3.4) which yields

$$s[p(s)] - [p(0)] = [a][p(s)] \tag{3.7}$$

with $[p(s)]$ being the L-transform of $[p(t)]$ and s designating the operator of this transform. $[p(0)]$ is the column vector of the initial probabilities of states. From (3.7) it follows that

$$[p(s)] = ([I_n]s-[a])^{-1}[p(0)] \tag{3.8}$$

where $[I_n]$ is the n by n identity matrix. The solutions are obtained by converting the expression for $[p(s)]$ into the time domain. The matrix inversion in (3.8), which should be performed in its general form, limits the application of the L-transform to systems having a very moderate number of states.

The solution of (3.4) may be expressed using the exponential matrix form

$$[p(t)] = \exp\{[a]t\} \, [p(0)] \tag{3.9}$$

For moderate t the exponential function in (3.9) may be approximated by the first q terms of the associated power series

$$[p(t)] \approx \left[\sum_{k=0}^{q} [a]^k \frac{t^k}{k!} \right] [p(0)], \qquad [a]^0 \equiv I_n \qquad (3.10)$$

How large q has to be can be roughly estimated by comparing the magnitude of the last term of the expansion with the entire sum.

The state probabilities may be assessed by subdividing the time interval under consideration into, say, r small intervals Δt. For small Δt we can write

$$\exp\{[a]r\Delta t\} = \left(\exp\{[a]\Delta t\}\right)^r \approx \left([I_n]+[a]\Delta t\right)^r \qquad (3.11)$$

Thus, with reference to (3.9)

$$[p(k\Delta t)] \approx \left([I_n]+[a]\Delta t\right)^k [p(0)], \qquad k=1,\dots,r \qquad (3.12)$$

It is clear from (3.12) that the following recurrent formula may be used

$$[p(k\Delta t)] \approx \left([I_n]+[a]\Delta t\right) [p((k-1)\Delta t)] \qquad (3.13)$$

To check whether Δt has been made small enough, it can be reduced to some extent and the calculations repeated. If the solutions obtained for the two values of Δt agree satisfactorily they can be taken to be good approximations.

Example 3.1

The state-transition diagram of a system composed of two two-state components is depicted in Fig. 3.1. It is presumed that the components are exposed to common-cause outages leading to the simultaneous outage of both components. These

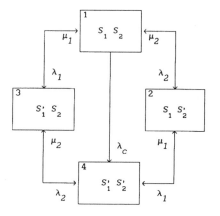

Fig. 3.1 State-transition diagram for a system exposed to common-cause failures

outages are usually caused by single events of external origin such as unfavorable weather conditions, mechanical shocks, environmental impacts, etc. S_k, $k=1,2$, denote the up states while $S_k{'}$ are the down states. λ_k and μ_k, $k=1,2$, are failure and renewal component transition rates. λ_c is the rate of transition due to common-cause failures. The diagram displayed might model two electric power transmission lines sharing the same right of way, exposed to the combined effects of wind and ice, landslide or earthquakes. A similar model may be applied to telecommunications lines, pipelines and many other systems consisting of two units sharing the same location. Equations (3.4) for this case are

$$
\begin{bmatrix} p_1'(t) \\ p_2'(t) \\ p_3'(t) \\ p_4'(t) \end{bmatrix} = [G] \begin{bmatrix} p_1(t) \\ p_2(t) \\ p_3(t) \\ p_4(t) \end{bmatrix}
$$

$$
[G] = \begin{bmatrix}
-(\lambda_1 + \lambda_2 + \lambda_c) & \mu_2 & \mu_1 & 0 \\
\lambda_2 & -(\lambda_1 + \mu_2) & 0 & \mu_1 \\
\lambda_1 & 0 & -(\mu_1 + \lambda_2) & \mu_2 \\
\lambda_c & \lambda_1 & \lambda_2 & -(\mu_1 + \mu_2)
\end{bmatrix}
$$

□

3.2
Composition of System Equations from Component Models

For composite systems consisting of many components the construction of system equations (3.4) is not trivial even for independent components because of the large number of system states and interstate transitions which should be generated. A computer oriented method for such a construction will be outlined in this section for both independent and dependent system components.

3.2.1
Independent Components

The Kolmogorov equations of a system built from independent components can be constructed by applying the Kronecker algebra [2,3,9].

The Kronecker product of an m by n matrix [a] and a p by q matrix [b] is an mp by nq matrix defined as

$$[a]\otimes[b] \equiv \begin{bmatrix} a_{11}[b] & \cdots & a_{1n}[b] \\ \cdot & \cdot & \cdot \\ a_{m1}[b] & \cdots & a_{mn}[b] \end{bmatrix} \tag{3.14}$$

with a_{ik} being elements of matrix [a].

The Kronecker sum of a q by q matrix [c] and an m by m matrix [d] is defined as

$$[c]\oplus[d] \equiv [c]\otimes[I_m] + [I_q]\otimes[d] \tag{3.15}$$

where $[I_k]$, $k=m,q,$ is the k by k identity matrix.

From definitions (3.14) and (3.15) it follows that

For [a] $\neq$ [b] :

$$[a]\otimes[b] \neq [b]\otimes[a] \tag{3.16}$$

$$[a]\oplus[b] \neq [b]\oplus[a]$$

$$[a]\otimes[b]\otimes[c] = [a]\otimes([b]\otimes[c]) = ([a]\otimes[b])\otimes[c]$$

$$[a]\oplus[b]\oplus[c] = [a]\oplus([b]\oplus[c]) = ([a]\oplus[b])\oplus[c] \tag{3.17}$$

$$([a]\otimes[b])([c]\otimes[d]) = ([a][c])\otimes([b][d])$$

The last equation in (3.17) implies that matrices [a] and [c] as well as [b] and [d] have dimensions allowing multiplication.

Let us assume that a system is built from n components with transition-rate matrices $[a_i]$, $i=1,...,n$. For independent components, which may, by definition, transit to any state whatever are the states of other system components, the system transition-rate matrix can be obtained as the Kronecker sum of matrices $[a_i]$

$$[a] = \bigoplus_{i=1}^{n} [a_i] \tag{3.18}$$

Define the state-event S_{ij}: component i is in state j. Then we can introduce the state-event column vector comprising all possible states of component i

$$[S_i] \equiv [S_{i1}\ S_{i2}\ ...\ S_{in_i}]^T \tag{3.19}$$

with n_i being the total number of states of component I.

The state-event vector of the system is obtainable [3] from the Kronecker product of vectors $[S_i]$

$$[S] = \bigotimes_{i=1}^{n} [S_i] \tag{3.20}$$

The sequence of components in (3.20) should be the same as in (3.18).

From (3.19) and (3.20) it is clear that there is a simple one to one correlation between the system state-event vector $[S]$ elements and the elements of the system state probability vector $[p(t)]$: $p_k(t)$ is the probability of the event in the kth row of $[S]$.

Example 3.2

Consider a system containing two independent components. The components both have two states: sound and failure states. The associated failure and renewal rates are λ_i and μ_i, $I=1,2$.

The component transition rate matrices are, with reference to (2.10) and (3.1),

$$[a_i] = \begin{bmatrix} -\lambda_i & \mu_i \\ \lambda_i & -\mu_i \end{bmatrix} \qquad i=1,2 \tag{3.21}$$

The state-event vectors are

$$[S_i] = [S_{i1} \ S_{i2}]^T \qquad i=1,2 \tag{3.22}$$

Indices 1 and 2 of elements in $[S_i]$ correspond to sound and failure states, respectively.

In accordance with (3.15), (3.18) and (3.22) we have

$$[a] = \begin{bmatrix} -\lambda_1 & \mu_1 \\ \lambda_1 & -\mu_1 \end{bmatrix} \otimes \begin{bmatrix} 1 & 0 \\ 0 & 1 \end{bmatrix} + \begin{bmatrix} 1 & 0 \\ 0 & 1 \end{bmatrix} \otimes \begin{bmatrix} -\lambda_2 & \mu_2 \\ \lambda_2 & -\mu_2 \end{bmatrix}$$

$$= \begin{bmatrix} -\lambda_1 & 0 & \mu_1 & 0 \\ 0 & -\lambda_1 & 0 & \mu_1 \\ \lambda_1 & 0 & -\mu_1 & 0 \\ 0 & \lambda_1 & 0 & -\mu_1 \end{bmatrix} + \begin{bmatrix} -\lambda_2 & \mu_2 & 0 & 0 \\ \lambda_2 & -\mu_2 & 0 & 0 \\ 0 & 0 & -\lambda_2 & \mu_2 \\ 0 & 0 & \lambda_2 & -\mu_2 \end{bmatrix} = \tag{3.23}$$

$$= \begin{bmatrix} -(\lambda_1+\lambda_2) & \mu_2 & \mu_1 & 0 \\ \lambda_2 & -(\lambda_1+\mu_2) & 0 & \mu_1 \\ \lambda_1 & 0 & -(\mu_1+\lambda_2) & \mu_2 \\ 0 & \lambda_1 & \lambda_2 & -(\mu_1+\mu_2) \end{bmatrix}$$

With reference to (3.20) and (3.22) the state-event system vector is

$$[S] = [S_{11}S_{21} \quad S_{11}S_{22} \quad S_{12}S_{21} \quad S_{12}S_{22}]^T \tag{3.24}$$

which means that in the first system state both components are good, in the second state component 1 is good and component 2 is under renewal, in the third state component 1 is under renewal and component 2 is good and in the forth state both components are under renewal. The elements of $[p(t)]$ are the probabilities of these system states, in the same order as listed in (3.24).

☐

3.2.2
Dependent Components

Many engineering systems comprise groups of dependent components. We shall mention some of the most common modes of dependence. These and some others will be discussed in more detail in Section 3.5.

A common-cause failure causes the simultaneous failure of several components. This should be modeled by modifying the Kolmogorov equations associated with sound states of these components as well as the equations associated with their simultaneous renewal state. As an illustration, compare the transition rate matrices in *Examples 3.1* and *3.2*. As components 1 and 2 are exposed to a common-cause failure, the diagonal element in the first row of the transition rate matrix, associated with system state 1, is in *Example 3.1* decreased by λ_c when compared with the independent case in *Example 3.2*. Also, in *Example 3.1* there is a transition rate λ_c from state 1 to state 4.

In some cases, due to inadequate personnel or/and equipment, only a single component can be repaired at a time. This affects the transition rates from system states associated with the simultaneous failure states of two or more components which will then be equal to the renewal transition rate of the component with the highest priority among the failed components. If, e.g., in *Example 3.1* and *Example 3.2* component 1 has the priority, transition rate μ_2 should be omitted from the forth column of the system transition rate matrix as component 2 cannot be repaired while component 1 is under repair.

The coincidences of some component states may be impossible by the nature of the system. The overlapping of preventive maintenance of several components is avoided in many engineering systems to preserve their availability beyond some critical limits. In many cases the failure of a component of a group puts out of service all other components of this group which, then, cannot fail. Hence, the simultaneous failures of the components belonging to this group are impossible. If this is the case, the associated equations for the independent case should be modified by omitting all impossible states. All columns and rows of the system transition rate matrix associated with the impossible states should be discarded.

3.3
Simplification of the System State Probability Equations

3.3.1
Systems with Independent Components

Taking account of (3.19) and (3.20) the probability that a system is in a certain state, say m, at time instant t, equals

$$p_m(t) = \Pr\{S_{1i}\ S_{2k}\ S_{3j}\ ...\} \tag{3.25}$$

with S_{pq} designating the event: component p is in state q at time instant t. Indices i, k, j, ... denote the states of components 1,2,3, ... in system state m. For independent components (3.25) converts into

$$p_m(t) = \Pr\{S_{1i}\}\ \Pr\{S_{2k}\}\ \Pr\{S_{3j}\}\ ... =$$
$$= p_{1i}(t)\ p_{2k}(t)\ p_{3j}(t)\ ... \tag{3.26}$$

where $p_{pq}(t)$ denotes the probability of component p being in its state q at time t. As can be observed from (3.26), the probability of a system being in a certain state is obtainable as the product of the probabilities of system components being in the states defining the corresponding system state. These probabilities can be determined by solving the state probability equations of components

$$[p_i'(t)] = [a_i][p_i(t)] \tag{3.27}$$

with $[p_i]$ representing the column vector of component i state probabilities and $[a_i]$ being the associated transition rate matrix. Eqs. (3.27) have a considerably lower order than the corresponding equations written for the system as a whole which reduces the computational efforts.

The suggested method may also be used in cases when only some of the components are independent. In this case, the independent and the remaining components are analyzed separately from one another. The final result for the system is obtained by multiplying the probabilities of the states of the independent and the remaining components to determine the probabilities of all possible system states.

3.3.2
Merging of States

By merging is meant the replacement of a set of states by a single state whose probability equals the sum of the probabilities of the replaced states. The merging is taken as valid if it does not affect the probability of the remaining states.

Let **G** be the set of states which are merged and **H** the set of remaining system states. The states in **G** are labeled as $1,...,m$ while the indices of the remaining system states are $m+1,...,m+n$. The system state probability equations can be written in the form

$$[p_g'(t)] = [a_{gg}][p_g(t)] + [a_{gh}][p_h(t)]$$

$$[p_h'(t)] = [a_{hg}][p_g(t)] + [a_{hh}][p_h(t0]$$

(3.28)

Indices g and h are related to the states in sets **G** and **H**.

The states in **G** may be merged if (3.28) can be transformed into the following form

$$[p_G'(t)] = a_{GG}p_G(t) + [a_{Gh}][p_h(t)]$$

$$[p_h'(t)] = [a_{hG}]p_G(t) + [a_{hh}][p_h(t)]$$

(3.29)

where

$$p_G(t) = [M][p_g(t)]$$

(3.30)

with [M] being the m-dimensional row vector of units. Clearly, (3.30) is the matrix form of summing the probabilities. To compose the equivalent reduced system the parameters a_{GG}, $[a_{Gh}]$ and $[a_{hG}]$ should be determined. The order of system (3.29) is $n+1$ which means that it is for $m-1$ lower than this of system (3.28).

The multiplication of the first equation in (3.28) by [M] from the l.h.s. yields

$$p_G'(t) = [M][a_{gg}][p_g(t)] + [M][a_{gh}][p_h(t)]$$

(3.31)

By comparing (3.29) and (3.31) we conclude that the following relationship should always hold

$$[a_{Gh}] = [M][a_{gh}]$$

(3.32)

A sufficient condition for merging is that the probabilities of merged states are proportional. If that is the case, we can write

$$[p_g(t)] = [e_g]p_G(t)$$

(3.33)

where $[e_g]$ is an n-dimensional column vector whose elements are positive numbers invariant in time and adding up to 1.

If (3.33) holds, then, with reference to (3.31) and (3.29)

$$a_{GG} = [M][a_{gg}][e_g]$$

$$[a_{hG}] = [a_{hg}][e_g]$$

(3.34)

Relationship (3.33) is not the necessary condition for merging. The first equation in (3.28) may also be converted into the first equation in (3.29) if matrix $[a_{gg}]$ satisfies the following relationship

$$[M][a_{gg}] = -a[M] \tag{3.35}$$

Then, according to (3.30)

$$a_{GG} = -a \tag{3.36}$$

The scalar form of (3.35) is

$$\sum_{i=1}^{m} a_{ik} = -a \qquad k=1,...,m \tag{3.37}$$

Bearing in mind that

$$a_{kk} = -\sum_{i=1}^{m+n} a_{ik} \qquad i \neq k \tag{3.38}$$

we deduce from (3.37) that

$$\sum_{i=m+1}^{m+n} a_{ik} = a \qquad k=1,...,m \tag{3.39}$$

Expression (3.39) is the alternative condition for forming the first equation in (3.29).

A sufficient condition for transforming the second equation in (3.28) into the second equation in (3.29) is, obviously,

$$[a_{hg}] = [b][M] \tag{3.40}$$

where [b] is an n-dimensional column vector whose elements b_i, $i=m+1,...,m+n$, are invariant in time.

The scalar form of (3.40) is

$$a_{ik} = b_i \qquad k=1,...,m, \quad i=m+1,...,m+n \tag{3.41}$$

From (3.41) it follows that the transition rates to state i in **H** from all states in **G** should be the same. This should be valid for all i in **H**, but the transition rates may differ for different i.

If (3.41) holds, then

$$\sum_{i=m+1}^{m+n} a_{ik} = \sum_{i=m+1}^{m+n} b_i \qquad k=1,...,m \tag{3.42}$$

The expression on the r.h.s. of (3.42) does not depend on k which means that if (3.40) holds, the same will be the case with (3.39) where

$$a = \sum_{i=m+1}^{m+n} b_i \qquad (3.43)$$

From the above it follows that (3.40) is an alternative single sufficient condition for merging.

Example 3.3

Consider a system consisting of two identical components exposed to common-cause failures having the state-transition diagram depicted in Fig. 3.2. The unit failure and renewal rates are λ and μ while the common-cause failure transition rate is λ_c.

From the symmetry of the state-transition diagram it can be concluded that states 1 and 2 are equally probable. Thus, we can write

$$\begin{bmatrix} p_1(t) \\ p_2(t) \end{bmatrix} = \begin{bmatrix} \dfrac{1}{2} \\ \dfrac{1}{2} \end{bmatrix} p_G(t)$$

with $p_G(t)$ being equal to the sum of $p_1(t)$ and $p_2(t)$. From (3.33) it follows that states 1 and 2 may be merged.

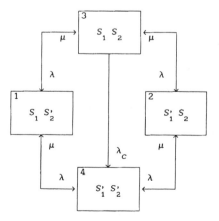

Fig. 3.2 System consisting of two identical components exposed to common cause failures

States 1 and 2 build set **G** and states 3 and 4 set **H**. Equations (3.28) for the system are

$$
\begin{bmatrix} [p_g'(t)] \\ [p_h'(t)] \end{bmatrix} =
\begin{bmatrix}
\begin{bmatrix} -(\lambda+\mu) & 0 \\ 0 & -(\lambda+\mu) \end{bmatrix} & \begin{bmatrix} \lambda & \mu \\ \lambda & \mu \end{bmatrix} \\
\begin{bmatrix} \mu & \mu \\ \lambda & \lambda \end{bmatrix} & \begin{bmatrix} -(2\lambda+\lambda_c) & 0 \\ \lambda_c & -2\mu \end{bmatrix}
\end{bmatrix}
\begin{bmatrix} [p_g(t)] \\ [p_h(t)] \end{bmatrix}
$$

With reference to (3.32) and (3.34) we determine

$$
a_{GG} = \begin{bmatrix} 1 & 1 \end{bmatrix}
\begin{bmatrix} -(\lambda+\mu) & 0 \\ 0 & -(\lambda+\mu) \end{bmatrix}
\begin{bmatrix} 1 & 1 \\ 2 & 2 \end{bmatrix} = -(\lambda+\mu)
$$

$$
[a_{hG}] = \begin{bmatrix} \mu & \mu \\ \lambda & \lambda \end{bmatrix}
\begin{bmatrix} 1 & 1 \\ 2 & 2 \end{bmatrix} = \begin{bmatrix} \mu \\ \lambda \end{bmatrix}
$$

$$
[a_{Gh}] = \begin{bmatrix} 1 & 1 \end{bmatrix}
\begin{bmatrix} \lambda & \mu \\ \lambda & \mu \end{bmatrix} = \begin{bmatrix} 2\lambda & 2\mu \end{bmatrix}
$$

Equations (3.29) are

$$
p_G'(t) = -(\lambda+\mu)p_G(t) + \begin{bmatrix} 2\lambda & 2\mu \end{bmatrix}
\begin{bmatrix} p_3(t) \\ p_4(t) \end{bmatrix}
$$

$$
\begin{bmatrix} p_3'(t) \\ p_4'(t) \end{bmatrix} = \begin{bmatrix} \mu \\ \lambda \end{bmatrix} p_G(t) +
\begin{bmatrix} -(2\lambda+\lambda_c) & 0 \\ \lambda_c & -2\mu \end{bmatrix}
\begin{bmatrix} p_3(t) \\ p_4(t) \end{bmatrix}
$$

The state-transition diagram of the merged system is displayed in Fig. 3.3.

Eqs. (3.29) could be deduced on the basis of the conservation of the state-transition frequency. That is, the total frequency of transitions from state 3 to states 1 and 2 equals, as observed from Fig. 3.2

$$
f_{3,12}(t) = \lambda p_3(t) + \lambda p_3(t) = 2\lambda p_3(t)
$$

To keep this frequency unchanged, the transition rate from state 3 to the state obtained by merging states 1 and 2 should be 2λ. Using the same reasoning we conclude that the transition rate from state 4 to the merged states should be 2μ.

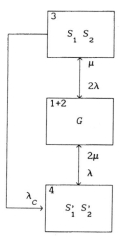

Fig. 3.3 State-transition diagram of the merged system

□

3.4
System Dependability Indices

3.4.1
System Availability

The first step in the dependability analysis of a system comprises the identification of possible system states and, then, the distinction between the up and down system states. Let us assume that there are v up states and w down states (Fig. 3.4). The

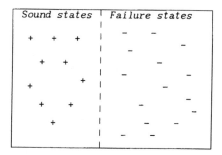

Fig. 3.4 System state space

system state probability equations can accordingly be partitioned

$$
\begin{bmatrix} [p_a'(t)] \\ [p_d'(t)] \end{bmatrix} = \begin{bmatrix} [a_{aa}] & [a_{ad}] \\ [a_{da}] & [a_{dd}] \end{bmatrix} \begin{bmatrix} [p_a(t)] \\ [p_d(t)] \end{bmatrix}
\tag{3.44}
$$

with indices a and d denoting the up and down states respectively. $[p_a(t)]$ is the v-dimensional column vector of the probabilities of the up system states while $[p_d(t)]$ is the w-dimensional column vector of the probabilities of the down system states. The transition rate matrix is partitioned into matrix blocks whose dimensions correspond to the dimensions of the associated system state probability vectors.

System availability equals to the total probability of the sound system states

$$
A(t) = [1_v][p_a(t)] = 1 - [1_w][p_d(t)]
\tag{3.45}
$$

System unavailability is

$$
U(t) = [1_w][p_d(t)] = 1 - [1_v][p_a(t)]
\tag{3.46}
$$

In (3.45) and (3.46) $[1_k]$, $k=v,w$, is a k-dimensional row vector of units. As is clear from (3.45) and (3.46), to calculate $A(t)$ and $U(t)$ we have to solve (3.44) for system state probabilities.

3.4.2
System Reliability and Mean Time to First Failure (MTFF)

System MTFF can be determined by following the same approach as used in Section 2.3. After the conversion of all system down states into absorbing states we have in (3.44)

$$
[a_{ad}] = [0]
$$

$$
[p_a'(t)] = [a_{aa}][p_a(t)]
\tag{3.47}
$$

The sum of the probabilities of the up system states deduced from (3.47) yields the probability that the system will be sound until time t, which is the system reliability, by definition.

With reference to (1.16)

$$
\text{MTFF} = \int_0^\infty \sum_i p_i(t)\,dt = \sum_i \lim_{s \to 0} p_i(s)
\tag{3.48}
$$

with $p_i(t)$ and $p_i(s)$ being the probability of system up state i and its L-transform. Index i runs over all system up states.

The L-transform of (3.47) is, after grouping of terms,

$$(s[I_v] - [a_{aa}]) [p_a(s)] = [p_a(0)] \tag{3.49}$$

with $[p_a(0)]$ designating the initial probabilities of system states. $[I_v]$ is the v by v identity matrix. From (3.49) we obtain

$$[a_{aa}] [\lim_{s \to 0} p_a(s)] = - [p_a(0)] \tag{3.50}$$

Eqn. (3.50) should be solved for $[\lim p_a(s)]$, $s \to 0$, and the solutions obtained inserted in (3.48).

Example 3.4

Let state 4 be the failure state of the system in *Example 3.3* . The MTFF should be determined under the assumption that at instant $t=0$ both units are up, i.e. $p_3(0)=1$.

The system state-transition diagram with state 4 being the absorbing state is depicted in Fig. 3.5.

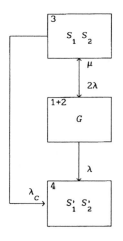

Fig. 3.5 Two-component system with absorbing state

For the system under consideration we have

$$[p_a'(t)] = \begin{bmatrix} p_G'(t) \\ p_3'(t) \end{bmatrix}$$

$$[a_{aa}] = \begin{bmatrix} -(\lambda+\mu) & 2\lambda \\ \mu & -(2\lambda+\lambda_c) \end{bmatrix}$$

Using (3.50) and (3.48) the following is obtained

$$\text{MTFF} = \frac{2\lambda+\lambda_c+\mu}{\lambda(2\lambda+\lambda_c) + \lambda_c\mu}$$

We leave it to the reader to prove this result.   □

3.5
Steady-State System Dependability Indices

As mentioned in Chapter 2, the relationships for the determination of the steady-state probabilities of system states are obtainable from (3.44) as limiting values for time t tending to infinity. Then

$$\lim_{t\to\infty} p_k'(t) = 0, \qquad \lim_{t\to\infty} p_k(t) = p_k = \text{const.} \tag{3.51}$$

with p_k being the steady-state probability of system state k, $k=1,...,v+w$. For steady state, (3.44) turns into a corresponding set of linear algebraic equations. It is implied that the system has no absorbing state. Otherwise, the steady-state probability of the absorbing state would be equal to 1 and all the other state probabilities would be zero.

The aforementioned set of linear equations is indefinite owing to (3.2). To make this system of equations solvable, an equation of the system should be replaced by the normalizing equation

$$\sum_{i=1}^{n} p_i = 1 \tag{3.52}$$

If the first equation is replaced, the following definite system is obtained

$$[A][p] = [b] \tag{3.53}$$

Matrix [A] is obtained from [a] by replacing each of its first row elements by unit. Column vector [b] has 1 as the first element while all its remaining elements are zeros.

3.5.1
System States Indices

Eqs. (3.53) are often called the *frequency balance equations* (Section 2.2) as for each system state k

$$- a_{kk} \, p_k = \sum_{i=1, \, i \neq k}^{n} a_{ki} \, p_i \qquad (3.54)$$

The l.h.s. term in (3.54) is the frequency of abandoning state k while the term on the r.h.s. yields the total frequency of encountering this state from all other system states.
  Parameter

$$\lambda_k \equiv - a_{kk} \qquad (3.55)$$

is the total transition rate of leaving state k, according to (3.2). It can be readily shown that the mean residence time in state k equals

$$m_k = \frac{1}{\lambda_k} \qquad (3.56)$$

The above may be proved using the method described in Section 3.2, by converting all system states except state k into absorbing states.
  The frequency of the transition from state k and the associated cycle duration are

$$f_k = \lambda_k p_k$$

$$T_k = \frac{1}{f_k} = m_k + r_k \qquad (3.57)$$

where r_k denotes the mean time spent outside state k. On the basis of (3.56) and (3.57) the following relationships are derived, correlating various dependability indices

$$p_k = f_k m_k$$

$$1 - p_k = f_k r_k \qquad (3.58)$$

$$r_k = \frac{1 - p_k}{f_k}$$

As can be observed, indices λ_k and m_k are the only indices which can be determined directly from the system transition rate matrix. To determine the rest of the indices, (3.53) should be solved for p_k .

The relationships derived show that an equivalent cyclic residence time diagram can be constructed for each system state, as was done for the two-state system discussed in Section 2.3.

3.5.2
System Indices

The partitioned system state probability Eqs. (3.44) written for the steady state are

$$
\begin{bmatrix} [A_{aa}] & [A_{ad}] \\ [A_{ad}] & [A_{dd}] \end{bmatrix} \begin{bmatrix} [p_a] \\ [p_d] \end{bmatrix} = [b] \tag{3.59}
$$

$[A_{ik}]$, i,k = a,d, are the sub matrices of $[A]$ having dimensions corresponding to $[p_a]$ and $[p_d]$.

The steady-state values of system availability and unavailability are determinable by inserting the steady-state probabilities obtained by solving (3.59) into (3.45) and (3.46).

System failure frequency equals the total frequency of the transitions from the up system states to the down system states. This frequency is the same as the total frequency of the transitions from the down system states to the up system states, as follows from the frequency balance property of the system steady state

$$
f = \sum_{i \in G, \, k \in H} a_{ki} \, p_i = \sum_{i \in H, \, k \in G} a_{ki} \, p_i \tag{3.60}
$$

In (3.60) **G** and **H** are sets of sound and down system states, respectively.
The system steady-state indices are

$$
\text{MTBF} = T = \frac{1}{f}
$$

$$
\text{MTTF} = m = \frac{A}{f} \tag{3.61}
$$

$$
\text{MTTR} = r = \frac{U}{f}
$$

As seen, to calculate MTTF and MTTR, indices A and U should be known which means that (3.59) should be solved for system state probabilities. It is important to note that MTTF and MTFF differ fundamentally for a system. MTFF depends on the initial state of the system while MTTF is a steady-state parameter characterizing average behavior of the system during a long run.

3.5.3
Truncation of States

In some cases the number of system states is very large which makes the dependability analysis nearly intractable. In such cases various approximate methods are applied. A widely used approximation method is the reduction of system state space by truncating the low probability states [4]. System states are partitioned into sets of probable and low probability states. This partition can be done regarding the number of low probability events which are associated with a state by bearing in mind that a failure state of a component has a probability which is usually several orders of magnitude smaller than the probability of the component sound state.

Let the first m system states be the probable states and the remaining n system states be the low probability states. By truncation, the system state probability vector is approximated in the following way

$$
\begin{bmatrix} [p_m] \\ [p_n] \end{bmatrix} \approx \begin{bmatrix} [p_m^*] \\ [0] \end{bmatrix}
\tag{3.62}
$$

with $[p_m^*]$ designating the approximate probabilities of the probable states as obtained after truncation. With reference to (3.62), Eqs. (3.53) reduce to

$$
[A_{mm}][p_m^*] = [b_m]
\tag{3.63}
$$

where $[A_{mm}]$ is built from the first m rows and columns of matrix $[A]$. Column vector $[b_m]$ is built from the first m elements of vector $[b]$. As matrix $[A_{mm}]$ is for n orders lower than matrix $[A]$, (3.63) can be much more easily solved than (3.53). The effects of the truncation upon the accuracy of the approximate results may be assessed by repeating the calculations after the inclusion of a certain number of the previously discarded states. If the difference between the results obtained in these two calculations is small enough, the first result may be taken as a good approximation. Clearly, the set of probable states has to include all system states of particular interest.

Example 3.5
Consider the system depicted in Fig. 3.6. For brevity, here and further on the states of system components are simply denoted by corresponding numerals. We presume that the failure of either of the two components causes the failure of the system and that, therefore, these failure states should be taken into consideration. If state 4 is discarded as being of lower probability order, (3.63) becomes

$$
\begin{bmatrix} 1 & 1 & 1 \\ \lambda_2 & -(\lambda_1+\mu_2) & 0 \\ \lambda_1 & 0 & -(\lambda_1+\mu_2) \end{bmatrix} \begin{bmatrix} p_1^* \\ p_2^* \\ p_3^* \end{bmatrix} = \begin{bmatrix} 1 \\ 0 \\ 0 \end{bmatrix}
\tag{3.64}
$$

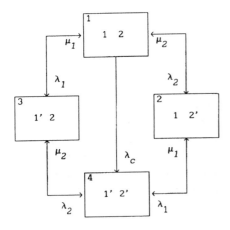

Fig. 3.6 Two-component system state-transition diagram

Table 3.1 gives the presumed components' dependability data. Table 3.2 lists the results of the approximate and exact calculation.

Table 3.1. Components' transition rates

$1/\lambda_1$	$1/\lambda_2$	$1/\lambda_c$	$1/\mu_1$	$1/\mu_2$
		h		
4380	3500	40 000	10	10

Table 3.2. Results of calculation

System state, i	p_i	p_i^*	Error, %
1	9.945×10^{-1}	9.949×10^{-1}	0.04
2	2.958×10^{-3}	2.835×10^{-3}	-4.1
3	2.395×10^{-3}	2.265×10^{-3}	-5.4
4	1.311×10^{-4}	0	–

As can be observed, the relative errors introduced in calculating the probabilities of states 2 and 3 are not very small and may not be ignored in some cases. This is because the discarded state 4 has a relatively high probability due to the common-cause failures implying a direct transition from state 1 to state 4.

□

3.5.4
Transition Rate Matrix Deviation Concept

Let us write (3.53) in the form [8]

$$[A_{gg}][p_g] + [A_{gh}][p_h] = [b]$$
$$[A_{hg}][p_g] + [A_{hh}][p_h] = [0] \tag{3.65}$$

We assume that $[A_{gg}]$ may be decomposed as

$$[A_{gg}] = [A_{gg}^*] + [d] \tag{3.66}$$

with $[A_{gg}^*]$ satisfying (3.67) and $[d]$ being a matrix of deviation from this matrix. We refer to the following system of equations

$$[A_{gg}^*][p_g^*] + [A_{gh}][p_h^*] = [b]$$
$$[A_{hg}][p_g^*] + [A_{hh}][p_h^*] = [0] \tag{3.67}$$

where probabilities $[p_i^*]$, $i=g,h$, are supposed to be known or easily determinable. It is assumed that Eqs. (3.67) are Eqs. (3.53) of a Markov system. Then it is clear from (3.66) that the elements of the first row of $[d]$ are zeros. The system described by (3.65) may be treated as generated from the system described by (3.67) by a deviation of the transition rate matrix. If the elements of $[d]$ are small enough, an approximate solution of (3.65) may be found in the form

$$[p_i] = [p_i^*] + [D_i] \qquad i=g,h \tag{3.68}$$

where $[D_i]$ is a column vector of relatively low norm, which has to be determined.

After inserting $[p_i]$ from (3.68) into (3.65) we obtain, with respect to (3.66) and (3.67)

$$[d] \, [p_g^*] + [A_{gg}][D_g] + [A_{gh}][D_h] = [0]$$
$$[A_{hg}][D_g] + [A_{hh}][D_h] = 0 \tag{3.69}$$

If $[d]$ is small, it can be taken that

$$[D_h] \approx [0] \tag{3.70}$$

as probabilities $[p_h]$ are not directly associated with the deviation matrix.

Taking account of (3.70) and (3.69) the following approximate relationship is derived

$$[A_{gg}][D_g] \approx - [d][p_g^*] \qquad (3.71)$$

If $[A_{gg}]$ is a matrix of low order, (3.71) can be easily solved for $[D_g]$. According to (3.68) and (3.71) the approximate solution of (3.65) is then

$$[p_g] \approx [p_g^*] + [D_g]$$
$$\qquad (3.72)$$
$$[p_h] \approx [p_h^*]$$

As the elements in the first row of $[A_{gg}]$ are units and the elements in the first row of $[d]$ are zeros, it follows from (3.71) that the sum of the elements of $[D_g]$ is zero. The sum of all elements of $[p_g^*]$ and $[p_h^*]$, taken together, equal to 1, because they have to obey the normalizing equation included in the first equation in (3.67). Consequently, the same is valid for the elements of $[p_g]$ and $[p_h]$ if calculated approximately, by applying (3.72).

Example 3.6
Consider the two-component system depicted in Fig. 3.7. Let us write the state probability equations of this system in the form of (3.65) by partitioning states 1 and 2 from states 3 and 4. If this is done, we have

$$[p_g] = \begin{bmatrix} p_1 \\ p_2 \end{bmatrix}, \qquad [p_h] = \begin{bmatrix} p_3 \\ p_4 \end{bmatrix},$$

$$[A_{gg}] = \begin{bmatrix} 1 & 1 \\ \lambda_c & -(\mu_1+\mu_2) \end{bmatrix}, \qquad [A_{gh}] = \begin{bmatrix} 1 & 1 \\ \lambda_2 & \lambda_1 \end{bmatrix},$$

$$[A_{hg}] = \begin{bmatrix} \lambda_1 & \mu_2 \\ \lambda_2 & \mu_1 \end{bmatrix}, \qquad [A_{hh}] = \begin{bmatrix} -(\mu_1+\lambda_2) & 0 \\ 0 & -(\mu_2+\lambda_1) \end{bmatrix}.$$

The state-transition diagram in Fig. 3.7 can be taken as generated by deviation from the associated diagram for two independent units with (see *Example 3.2*)

$$[A_{gg}^*] = \begin{bmatrix} 1 & 1 \\ 0 & -(\mu_1+\mu_2) \end{bmatrix}$$

The deviation matrix is

$$[d] = \begin{bmatrix} 0 & 0 \\ \lambda_c & 0 \end{bmatrix}$$

The steady-state solutions for the generating system are trivial

$$p_1^* = A_1 A_2 , \quad p_2^* = U_1 U_2 , \quad p_3^* = U_1 A_2 , \quad p_4^* = A_1 U_2$$

where

$$A_k = \frac{\mu_k}{\lambda_k + \mu_k} , \quad U_k = \frac{\lambda_k}{\lambda_k + \mu_k} , \quad k=1,2$$

By solving (3.71) with $[A_{gg}]$ and p_1^* and p_2^* as deduced above, we obtain

$$D_1 \approx - \frac{\lambda_c}{\lambda_c + \mu_1 + \mu_2} p_1^*$$

$$D_2 \approx \frac{\lambda_c}{\lambda_c + \mu_1 + \mu_2} p_1^*$$

Finally, the approximate solutions for the system under consideration are, in accordance with (3.72)

$$P_1 \approx p_1^* - \frac{\lambda_c}{\lambda_c + \mu_1 + \mu_2} p_1^*$$

$$P_2 \approx p_2^* + \frac{\lambda_c}{\lambda_c + \mu_1 + \mu_2} p_1^*$$

$$P_3 \approx p_3^* , \quad P_4 \approx p_4^*$$

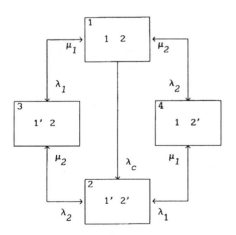

Fig. 3.7 System state-transition diagram

It is important to note that by using the procedure described, explicit expressions are derived for all state probabilities. This can be helpful in analyzing the effects of various parameters upon the system dependability indices.

Table 3.3 gives the results of a comparative calculation conducted by applying the approximate and exact approaches. The same data for system components are adopted as in Table 3.1.

As may be observed from Table 3.3, the approximate method yields nearly exact results for the probabilities of states 1 and 2. These two states are usually the most important in practice. The errors introduced for states 3 and 4 are much higher and could not be tolerated in some cases if these two states are of prime importance.

Table 3.3 Results of calculation

System state, i	p_i^*	p_i exact	p_i approximate	Error %
1	9.949×10^{-1}	9.945×10^{-1}	9.947×10^{-1}	– 0.02
2	6.490×10^{-6}	1.311×10^{-4}	1.315×10^{-4}	– 0.30
3	2.271×10^{-3}	2.395×10^{-3}	2.271×10^{-3}	5.46
4	2.842×10^{-3}	2.958×10^{-3}	2.842×10^{-3}	4.08

□

3.5.5
Matrix Gauss–Seidel Approach

The Gauss–Seidel iterative approach for solving the systems of linear algebraic equations can be structured to operate with sets of unknown probabilities instead of with probabilities individually [5,6]. Such an procedure will be called the *matrix Gauss–Seidel approach* to emphasize the aforementioned modification of the general concept of the method. The matrix approach proves to be favorable when applied to Markov systems as it uses some properties of these systems to speed up the calculation procedure.

The state probability vector in (3.53) should be partitioned into subvectors encompassing the probabilities assessed to be of the same order of magnitude

$$[p] = [\ [P_1]^T \ [P_2]^T \ \dots \ [P_N]^T \ ] \tag{3.73}$$

Here $[P_k]$, $k=1,\dots,N$, is the m_k-dimensional vector of the system state probabilities belonging to probability set k. It is presumed that this set comprises m_k system states.

The sets of states are ordered from more probable to less probable states. Thus, the index of the most probable states is 1 while N is the index of the least probable states. The order of magnitude of a system state can be assessed by the number of low probability events associated with this state. The higher this number is the lower is the probability of the state under consideration. The failure or preventive maintenance component states are low probability states in most engineering systems, for illustration. Therefore, a system state associated with, say m, components being under renewal or maintenance has a higher magnitude probability order than a state associated with $m+1$ components in renewal or maintenance states.

If the system states are partitioned into sets as in (3.73), Eqs. (3.53) become

$$\begin{bmatrix} [A_{11}] & \cdots & [A_{1N}] \\ \cdot & & \cdot \\ \cdot & & \cdot \\ \cdot & & \cdot \\ [A_{N1}] & \cdots & [A_{NN}] \end{bmatrix} \begin{bmatrix} [P_1] \\ \cdot \\ \cdot \\ \cdot \\ [P_N] \end{bmatrix} = \begin{bmatrix} [b_1] \\ [0] \end{bmatrix} \tag{3.74}$$

$[A_{ik}]$ are m_i by m_k sub matrices obtained by the partition of $[A]$ with regard to (3.73). Column vector $[b_1]$ is built from the first m_1 elements of vector $[b]$ in (3.53).

In the initial iteration step, the "zero order iteration" of system state probabilities is calculated. The procedure begins with the determination of the zero iteration of the set of the most probable state probabilities, denoted as $[P_1^{(0)}]$. These probabilities are obtained from (3.74) after inserting zero values for all other, lower magnitude order probabilities

$$[A_{11}][P_1^{(0)}] = [b_1] \tag{3.75}$$

The zero order iteration for $[P_2]$ is calculated from (3.74) by discarding all lower magnitude order probabilities and by inserting the zero order iteration for $[P_1]$ determined in the preceding step. Hence, the equation for the calculation of $[P_2^{(0)}]$ is

$$[A_{21}][P_1^{(0)}] + [A_{11}][P_2^{(0)}] = [0] \tag{3.76}$$

Generally, the zero order iteration for probabilities $[P_k]$ is obtainable from (3.74) by discarding all lower magnitude order probabilities and inserting the zero order iteration for all higher magnitude order probabilities, deduced in the preceding steps. Hence, the equation for the calculation of $[P_k^{(0)}]$ is

$$\sum_{i=1}^{k-1} [A_{ki}][P_i^{(0)}] + [A_{kk}][[P_k^{(0)}] = [0] \tag{3.77}$$

By consecutively solving (3.77) for $[P_k^{(0)}]$, $k=2,...,N$, the zero iterations for all state probability vectors are obtained.

The first order iteration of $[P_k]$ is calculated from (3.74) by inserting first order iterations for all higher magnitude order probabilities and zero iterations for all lower magnitude order probabilities. Then the corresponding equation for $[P_k^{(1)}]$, $k=2,...,N$, is

$$\sum_{i=1}^{k-1} [A_{ki}][P_i^{(1)}] + [A_{kk}][P_k^{(1)}] + \sum_{i=k+1}^{N} [A_{ki}][P_i^{(0)}] = [0] \tag{3.78}$$

$[P_k^{(1)}]$ are the only unknowns in (3.78) as all other probabilities are calculated in the preceding calculation steps.

From (3.78) it could be concluded that the general expressions for the calculation of the qth iteration for probabilities $[P_k]$ are

$$[P_1^{(q)}] = [A_{11}]^{-1}[b_1] - \sum_{i=2}^{N} [A_{11}]^{-1}[A_{1i}][P_i^{(q-1)}]$$

$$[P_k^{(q)}] = -\sum_{i=1}^{k-1} [A_{kk}]^{-1}[A_{ki}][P_i^{(q)}] - \tag{3.79}$$

$$- \sum_{i=k+1}^{N} [A_{kk}]^{-1}[A_{ki}][P_i^{(q-1)}]$$

The matrix products in (3.79) are to be determined only once as they do not change during the calculation procedure. Eqs. (3.79) are general as they hold for all iterations and for all state probability vectors. The previously displayed equations are obtainable from (3.79) as special cases.

As stressed previously, the orders of magnitude of state probabilities can be easily assessed by counting the number of low probability events associated with a system state. There is another very favorable property of many engineering systems modeled by Markov processes. States with probabilities of the same order of magnitude do not communicate with each other. Transitions between states are commonly associated with the occurrence of failures or the termination of renewal or preventive maintenance. The first mentioned event causes the transition to a state of lower order of magnitude of probability while the second mentioned events lead to a transition to a state of higher order of magnitude of probability. Consequently, the state under consideration can transit to states of lower or higher order of magnitude of probability only, which explains the above statement. Owing to this, matrices $[A_{kk}]$ have a diagonal form whose inversion is trivial.

In many cases $[P_1^{(1)}]$ and $[P_k^{(0)}]$, $k=2,...,N$, are very close to the exact solutions and may be used as good estimates. This is particularly true when the orders of magnitude of state probabilities differ significantly. In this case further simplification may be introduced by omitting in (3.78) all transition rates to states of lower order of magnitude of probability.

Example 3.7

The matrix Gauss–Seidel procedure is to be used to determine the steady-state probabilities of states of the system in *Example 3.6*. The states are labeled as in Fig. 3.6. Inspection of the state transition-diagram in Fig. 3.6 leads to the conclusion that the system probabilities may be partitioned into three sets: the set of the most probable states contains state 1, the second most probable state set comprises states 2 and 3 and the least probable state set is represented by state 4. Consequently,

$$[P_1] = [p_1] , \qquad [P_2] = \begin{bmatrix} [p_2] \\ [p_3] \end{bmatrix} , \qquad [P_3] = [p_4]$$

The sub-matrices in (3.74) for the system under consideration are

$$[A_{11}] = [1] , \qquad [A_{12}] = [1 \;\; 1] , \qquad [A_{13}] = [1] ,$$

$$[A_{21}] = \begin{bmatrix} \lambda_2 \\ \lambda_1 \end{bmatrix} , \qquad [A_{22}] = \begin{bmatrix} -(\lambda_1 + \mu_2) & 0 \\ 0 & -(\lambda_2 + \mu_1) \end{bmatrix} , \qquad [A_{23}] = \begin{bmatrix} \mu_1 \\ \mu_2 \end{bmatrix}$$

$$[A_{31}] = [\lambda_c] , \qquad [A_{32}] = [\lambda_1 \;\; \lambda_2] , \qquad [A_{33}] = [-(\mu_1 + \mu_2)] .$$

Table 3.4 shows the results of the iteration procedure.

Table 3.4. Results of calculation

q	0	1	2	exact
$p_1^{(q)}$	1	0.9947	0.9945	0.9945
$p_2^{(q)}$	2.85×10^{-3}	2.96×10^{-3}	2.97×10^{-3}	2.96×10^{-3}
$p_3^{(q)}$	2.27×10^{-3}	2.39×10^{-3}	2.39×10^{-3}	2.39×10^{-3}
$p_4^{(q)}$	1.29×10^{-4}	1.31×10^{-4}	1.31×10^{-4}	1.31×10^{-4}

Obviously, the iterations rapidly converge to the exact solutions. As can be observed, the first order iterations are very close to the exact results.

□

3.6
Characteristic Applications

3.6.1
Independent Components

The state-transition diagram of a system composed from two two-state units is displayed in Fig. 3.8. This diagram, as found from previous discussions, enables the calculation of the probabilities of all system states. However, to determine the dependability indices of this system we have to know which among the four system states are sound states. If the sound operation of both units is the condition for functioning of the system, state 1 is the only good system state. If this is the case, the units are said to be *series connected* in the logical sense. For the systems that fail only if both units are down, state 4 will be the only failure state. Such systems are *redundant* with *parallel connected* units, in the logical sense. Clearly, many practical systems are physically built as series and parallel constructions. From the above it is obvious that the state- transition diagram in Fig. 3.8 may model both types of systems, which only differ by the boundary lines separating the sound and failure states. These lines are indicated in Fig. 3.8. The unit failure and renewal transition rates are λ_k and μ_k, $k=1,2$.

As the units are independent, the probabilities of system states are obtainable by multiplying the probabilities of the corresponding unit states. Consequently, we have

$$p_1(t) = A_1(t)A_2(t) \ , \quad p_2(t) = A_1(t)U_2(t) \ ,$$

$$p_3(t) = U_1(t)A_2(t) \ , \quad p_4(t) = U_1(t)U_2(t) \tag{3.80}$$

with $A_k(t)$ and $U_k(t)$, $k=1,2$, being the availability and unavailability of unit k, as determined in Section 2.2. Clearly, (3.80) holds for the steady-state probabilities too.

For the series connection, the system availability, unavailability and failure transition rate are, regarding the boundary line in Fig. 3.8,

$$A(t) = A_1(t)A_2(t) \ ,$$

$$U(t) = A_1(t)U_2(t) + U_1(t)A_2(t) + U_1(t)U_2(t)$$

$$= U_1(t)+U_2(t)-U_1(t)U_2(t) = 1-A_1(t)A_2(t) \tag{3.81}$$

$$\lambda = \lambda_1 + \lambda_2$$

The steady-state system failure frequency equals the frequency of leaving state 1. Thus,

$$f = (\lambda_1+\lambda_2)A = \lambda_1 A_1 A_2 + \lambda_2 A_2 A_1 = f_1 A_2 + f_2 A_1 \tag{3.82}$$

with f_k , $k=1,2$, denoting the failure frequency of unit k.

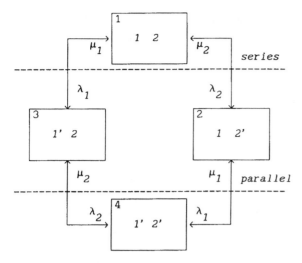

Fig. 3.8 State-transition diagram for the system with two independent components

As mentioned before, for the redundant parallel connection, state 4 is the only failure state. Consequently, for this case

$$U(t) = U_1(t)U_2(t)$$

$$A(t) = U_1(t)A_2(t)+A_1(t)U_2(t)+A_1(t)A_2(t)$$

$$= A_1(t)+A_2(t)-A_1(t)A_2(t) = 1-U_1(t)U_2(t)$$

(3.83)

$$\mu = \mu_1 + \mu_2$$

where μ is the system renewal rate. It should be stressed that unit availabilities are often very close to unity. Therefore, to avoid errors due to rounding, the last expression for $A(t)$ in (3.83) is recommended.

The steady-state system failure frequency equals the frequency of leaving the failure state. Hence,

$$f = (\mu_1+\mu_2)U = \mu_1 U_1 U_2 + \mu_2 U_2 U_1 = f_1 U_2 + f_2 U_1$$

(3.84)

If we compare the expressions derived for the series connected and the parallel connected redundant system case we conclude that these two systems can be taken as *dual systems*. The dependability indices of the series system are obtainable from the expressions for the parallel redundant system by replacing availabilities by unavailabilities, failure transition rates by renewal transition rates, and vice versa.

If a system is composed of two parallel connected units which are not redundant, state 1 is the only sound system state. Consequently, for such a system the expressions deduced for the series system are valid.

3.6.2
Preventive Maintenance

Let us consider a two-unit redundant parallel connected system. The units may be in the sound, failure and preventive maintenance states. The *preventive maintenance* should be conducted by following some rules prescribed to improve the availability of the system. The maintenance of a unit should not begin if the other unit has failed and is under repair. Also, the simultaneous maintenance of units is not allowed. The state-transition diagram for the system under consideration is depicted in Fig. 3.9. The transition rates to and from the maintenance state are marked by a double quote. As can be deduced, states 6, 7 and 8 are system down states. In accordance with the rules adopted, the state associated with the simultaneous maintenance of units is missing. Also, there are no transitions from state 2 to state 7 and from state 3 to state 8 which would conflict with the rule preventing the beginning of the maintenance of a unit while the other unit is down.

Obviously, the units of the system are dependent as their transitions to preventive maintenance states are conditioned by the status of the other unit of the pair. Therefore, it is not possible to obtain the time-specific probabilities of system states

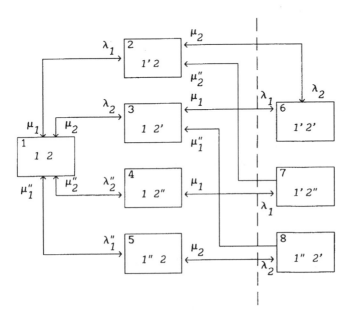

Fig. 3.9 State-transition diagram for the two-unit system with preventive maintenance

in a general closed form but they should be calculated by solving the associated Kolmogorov differential equations composed by referring to the diagram in Fig. 3.9. The steady-state solutions could be obtained in closed form by solving the corresponding system of linear equations as discussed in Section 3.5. However, the solutions obtained would be very cumbersome owing to the high order of the system of equations. Here, we shall derive the expressions for the basic dependability indices of the system by applying the approximate matrix Gauss–Seidel approach.

The most probable is state 1, the second most probable are states 2, 3, 4 and 5, while states 6, 7 and 8 constitute the set of least probable states. By bearing in mind the above classification of states we can write

$$p_1 \approx 1 \ , \quad \mu_1 p_2 - \lambda_1 \approx 0 \ , \quad \mu_2 p_3 - \lambda_2 \approx 0$$

$$\mu_2'' p_4 - \lambda_2'' \approx 0 \ , \quad \mu_1'' p_5 - \lambda_1'' \approx 0 \ , \tag{3.85}$$

$$\mu_1 p_6 - \lambda_2 p_2 - \lambda_1 p_3 \approx 0 \ , \quad \mu_2'' p_7 - \lambda_1 p_4 \approx 0 \ , \quad \mu_1'' p_8 - \lambda_2 p_5$$

By solving (3.85) we obtain

$$p_2 \approx \frac{\lambda_1}{\mu_1} \approx U_1 \ , \quad p_3 \approx \frac{\lambda_2}{\mu_2} \approx U_2 \ ,$$

$$p_4 \approx \frac{\lambda_2''}{\mu_2''} \approx U_2'' \ , \quad p_5 \approx \frac{\lambda_1''}{\mu_1''} \approx U_1'', \tag{3.86}$$

$$p_6 \approx U_1 U_2 \ ,$$

$$p_7 \approx U_1 U_2'' K_{12''} \ , \qquad p_8 \approx U_1'' U_2 K_{21''}$$

where

$$U_i'' \equiv \frac{\lambda_i''}{\lambda_i'' + \mu_i''} \ , \quad K_{ij''} \equiv \frac{\mu_i}{\mu_i + \mu_j''} \ , \quad i, j = 1, 2, \ i \neq j \tag{3.87}$$

The dependability indices of the system are, with regard to (3.85) and (3.86)

$$U \approx U_1 U_2 + U_1'' U_2 K_{21''} + U_2'' U_1 K_{12''}$$

$$f \approx (\mu_1 + \mu_2) U_1 U_2 + \mu_1 U_1 U_2'' + \mu_2 U_1'' U_2 \tag{3.88}$$

$$= f_1 U_2 + f_2 U_2 + f_1 U_2'' + f_2 U_1''$$

3.6.3
Restricted Repair

Consider a redundant system composed of two parallel connected two-state units which is serviced by a single repair crew. For this reason, the simultaneous repair of both units is impossible. The rule is: if both units are down, the unit which failed first should be repaired first. The state-transition diagram for such a system is displayed in Fig. 3.10 [7]. Symbols k_w and k' denote "unit k is waiting for repair" and "unit k is under repair", respectively. Owing to the restricted repair facilities, the transitions from state 4 to state 2 and from state 5 to state 3 are not possible. States 4 and 5 are system down states.

By applying the approximate Gauss–Seidel method, the following solutions for the steady-state dependability indices of the system are derived

$$U = p_4 + p_5 \approx \lambda_1\lambda_2(\mu_1^{-2}+\mu_2^{-2})$$

$$f = \mu_1 p_4 + \mu_2 p_5 \approx \lambda_1\lambda_2\frac{\mu_1\mu_2}{\mu_1+\mu_2}$$

(3.89)

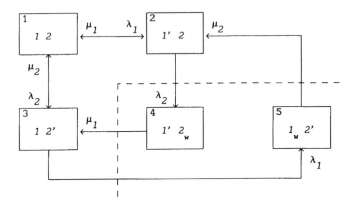

Fig. 3.10 State-transition diagram for the case of restricted repair

3.6.4
Partially Redundant Systems

Let us consider a system consisting of n identical two-state units which is operable if at most m of these units are nonoperable. Such systems are also known as m out of n: *Good* systems or *majority voting systems*

It is clear that the reliability of such systems equals

$$R(t) = \sum_{k=0}^{m} \binom{n}{k} Q_u(t)^k R_u(t)^{n-k} \tag{3.90}$$

with $R_u(t)$ and $Q_u(t)$ being the reliability and unreliability of a single unit. The sum in (3.90) covers all possible coincidences of the up and down states of units for which the system is operable. For $k=1$, for example, (3.90) yields the sum of the probabilities of all system states in which a unit has failed before t while all remaining units are operable. The number of these states is n as any among n units may fail before t. For $k=2$ all combinations of two failed units among n units are summed, etc. Good system states are for all $k \le m$.

With reference to (1.16) and (3.90) we have

$$\text{MTFF} = \sum_{k=0}^{m} \binom{n}{k} \int_{0}^{\infty} Q_u(t)^k R_u(t)^{n-k} dt \tag{3.91}$$

System availability is, in analogy to (3.90),

$$A(t) = \sum_{k=0}^{m} \binom{n}{k} U_u(t)^k A_u(t)^{n-k} \tag{3.92}$$

where $A_u(t)$ and $U_u(t)$ denote the availability and unavailability of a single unit.

The steady-state failure frequency of the system equals the frequency of the transition from the states with m nonoperable units to the state with $m+1$ nonoperable units as these two sets of states are boundary system up and down states, respectively. Consequently,

$$\begin{aligned}
f &= \binom{n}{m} U_u^m A_u^{n-m} (n-m)\lambda \\
&= \binom{n}{m+1} U_u^{m+1} A_u^{n-m-1} (m+1)\mu
\end{aligned} \tag{3.93}$$

where λ and μ are the failure and renewal transition rates of a unit.

3.6.5
Induced failures

In many cases in practice a failure of a unit may cause damage to an adjacent unit if this failure is associated with large dissipation of heath or an explosion. The state transition diagram for such a two-unit system is depicted in Fig. 3.11 with α_k, $k=1,2$, denoting the probability that failure of unit k will induce the failure of the other

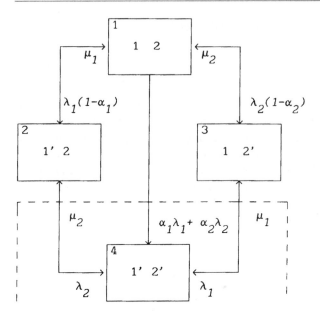

Fig. 3.11 State-transition diagram for a two-unit system with induced failures

system unit. By applying the approximate Gauss–Seidel method the following expressions for the steady-state probabilities of system states are deduced

$$p_2 \approx (1-\alpha_1)\frac{\lambda_1}{\mu_1}, \qquad p_3 \approx (1-\alpha_2)\frac{\lambda_2}{\mu_2},$$

$$p_4 \approx \frac{\alpha_1\lambda_1 + \alpha_2\lambda_2 + \lambda_2 p_2 + \lambda_1 p_3}{\mu_1 + \mu_2}$$

(3.94)

For a redundant system the steady-state system unavailability equals p_4 and the failure frequency equals the frequency of encountering state 4

$$f = p_4(\mu_1 + \mu_2) = \alpha_1\lambda_1 + \alpha_2\lambda_2 + \lambda_2 p_2 + \lambda_1 p_3$$

(3.95)

3.6.6
System with a Cold Standby Unit

Consider a system consting of unit 1 which is in service and a cold standby unit 2. When unit 1 fails, unit 2 is set into service by an imperfect switch which might stick with probability α. The mean switching time s is short but finite. After renewal, unit 1 is set in service and unit 2 is reset into the standby mode. It is presumed that unit

2 is not exposed to failures while residing in the standby mode. The state transition diagram for the system under consideration is displayed in Fig. 3.12. Index s indicates the standby mode.

The total rate of transitions from state 2 is $1/s$ while the transition rates to states 3 and 5 are the expected values concerning the switch sticking probability α. Obviously, states 2 and 5 are the failure states for the system. The steady-state dependability indices of the system can again be assessed by applying the approximate Gauss–Seidel method. The most probable is state 1. States 2 and 3 constitute the set of the second most probable states while the remaining two states form the set of the least probable states. It should be noted that state 4 is encountered exclusively from the low probability state 5 which makes this state also of low probability. From the aforesaid, the following approximate relationships are obtained

$$p_1 \approx 1$$

$$-\frac{1}{s}p_2 + \lambda_1 \approx 0$$

$$-\mu_1 p_3 + (1-\alpha)\frac{1}{s}p_2 \approx 0 \tag{3.96}$$

$$-(\mu_2+\lambda_1)p_4 + \mu_1 p_5 \approx \qquad -\mu_2 p_4 + \mu_1 p_5 \approx 0$$

$$-(\mu_1+\mu_2)p_5 + \frac{\alpha}{s}p_2 + \lambda_2 p_3 + \lambda_1 p_4 \approx 0$$

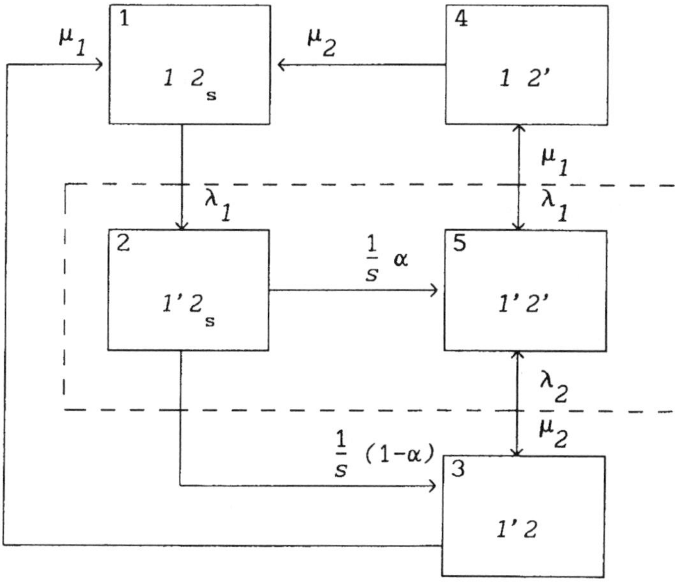

Fig. 3.12 State-transition diagram for a two-unit system with an unit in cold standby mode

System unavailability equals, as deduced from (3.96)

$$U = p_2 + p_5 \approx s\lambda_1 + \frac{\alpha\lambda_1}{\mu_1 + \mu_2} + \frac{(1-\alpha)\lambda_1\lambda_2}{\mu_1(\mu_1 + \mu_2)} \tag{3.97}$$

The first term in the second expression in (3.97) is the approximation of p_2 while the remaining two terms yield the approximate value of p_5. The first term is usually much greater than the other two.

The failure frequency can be determined as the total frequency of the transitions from states 2 and 5 to good system states

$$f = \frac{1-\alpha}{s}p_2 + (\mu_1 + \mu_2)p_5 \approx \lambda_1 + (1-\alpha)\frac{\lambda_1\lambda_2}{\mu_1} \tag{3.98}$$

The first term is most commonly prevailing.

Problems

1. Construct the state-transition diagram of a system consisting of three two-state units. The system is up if at least two units are sound. If two units are down the system is not operating and the sound unit left cannot fail. The system is served by a single repair crew which applies the rule: first failed first repaired. (Refer to Section 3.6 for various examples)

2. Presume that the units of the system in *Problem 1* have the same dependability indices. Simplify the system state-transition diagram by merging the states. (Refer to Section 3.5)

3. Consider three series connected two-state units modeling a transmission system. If any unit is down the remaining units cannot fail. Construct the state-transition diagram for this system and determine the system steady-state unavailability and failure frequency. (Refer to section 3.6)

4. Determine the MTFF for the system in *Problem 1* if all units are sound at $t = 0$. (Refer to Section 3.4)

5. Find the approximate expressions for the steady-state probabilities of the states of the system in *Problem 1* by applying the simplified Gauss–Seidel approach. (Refer to Sections 3.5 and 3.6)

References

1. N.S. Bakhvalov, *Numerical Methods,* Mir Publishers, Moscow (1977)
2. Cafaro, G., Corsi, F., Vacca, F. Multistate Markov models and structural properties of the transition-rate matrix, *IEEE Trans. Reliab.,* **R-35** (1986), pp. 192–200.
3. Nahman, J., Comment on: 'Multistate Markov models and structural properties of the transition-rate matrix', *IEEE Trans. Reliab.,* **R-36** (1987), pp. 639–640.
4. Singh, C., Billinton, R., *System Reliability Modelling and Evaluation,* Hutchinson & Co., London (1977)
5. Nahman, J., Approximate expressions for steady-state reliability indices of Markov systems, *IEEE Trans. Reliab.,* **R-35** (1986), pp. 338–343.
6. Nahman, J., Iterative method for steady-state reliability analysis of complex Markov systems, *IEEE Trans. Reliab.,* **R-33** (1984), pp. 406–409.
7. Billinton, R., Alam, M., Effect of restricted repair on system reliability indices, *IEEE Trans. Reliab.,* **R-27** (1978), pp. 376–379.
8. Nahman, J., Approximate steady-state solutions for Markov systems based upon the transition-rate matrix deviation concept, *Microelectr. Reliab.,* **34** (1994), pp. 7–15.
9. Nahman, J., Graovac, M., A method for evaluating the frequency of deficiency-states of electric-power systems, *IEEE Trans. Reliab.,* **R-39** (1990), pp. 265–272.

4 Networks

The dependability of many engineering systems can be analyzed using adequate network structures. The interruption of system functioning is modeled by the interruption of the connection between two characteristic nodes of the associated network, usually named *the source* and *sink (terminal) node*. The components of the system are represented as network branches. The failure state of a component is modeled as an interruption of the corresponding branch. The network modeling of a system is alternatively colled a *logic diagram* or *reliability block diagram*. We shall use, further on, the term *function graph* as it clearly reflects the purpose of a network model. The analysis exclusively uses the graph of the corresponding network as it completely reflects its structure relevant for dependability evaluation. The graphs appearing in our analysis may generally contain parallel branches which means that more precisely speaking they are multigraphs. For brevity, the term graph is used further on also for graphs with parallel branches.

An engineering system may be represented by a function graph if it possesses the following properties [1,2]:

a) The system is composed of two-state components. The states are: sound and failed.
b) The system has also only two states: sound and failed.
c) The renewal of a component cannot worsen and the failure of a component cannot improve the functional ability of the system.

Systems having property c) are named *monotonous* or *coherent* systems.

The function graph models the impacts of failures upon system operation and it need not necessarily reflect the physical structure of a system. Consider, for example, a system consisting of three units A, B and C. If the function of the system is interrupted by the failure of any one of these units, the function graph of the system should be that depicted in Fig. 4.1 *a)*. If two sound units suffice for successful system operation, the corresponding function graph is that in Fig. 4.1 *b)*. It indicates that the system is in the failure state if any two units are simultaneously nonoperable. The function graph in Fig 4.1 *c)* is valid for the case when the system is completely redundant, i.e., for a satisfactorily functioning system it is sufficient that any of the system units is sound. The associated graph indicates that the system is down only if all three units are simultaneously down, that is, the connection between the source and sink node will be interrupted if all graph branches are interrupted.

The rules for composing a function graph are basically simple. The interruption of the connection between the source and the sink node models the system down state, as mentioned before. Branches are parallel connected if the system is down

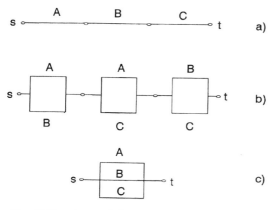

Fig. 4.1 Function graph for a system with three units

when all corresponding system components are down. All events causing the down state of the system are series connected. However, for composite systems the construction of the function graph can be a very serious task as it implies the identification of all system down states and associated failure events. Fortunately, there are many important engineering systems which are physically constructed and operated as networks and whose structural graphs are identical or closely related to the function graph used in the dependability analysis. Examples include electrical power transmission and distribution networks, telecommunications and computer networks, water supply networks, and traffic networks.

Network models are a relatively simple means for deducing the most critical events and associated dependability indices of engineering systems which makes them preferable to some other more detailed but laborious system models, in many cases.

4.1
Elementary Network Structures

Function graphs are, in many cases, composed of series and/or parallel connected mutually independent branches, in various structural forms. For these elementary network structures, the system dependability indices can be deduced in closed analytical form which provides a clear insight into the criticality of various events and parameters.

4.1.1
Series Connection

Let a function graph be constituted of n series connected branches. This system is sound only if all its components are sound, i.e., if none of the function graph branches is interrupted. The system reliability is consequently

$$R(t) = \prod_{i=1}^{n} R_i(t) \tag{4.1}$$

with $R_i(t)$ being the reliability of component i. The unreliability of the system is then, with regard to (1.3),

$$Q(t) = 1 - \prod_{i=1}^{n} R_i(t) = 1 - \prod_{i=1}^{n} (1-Q_i(t)) \tag{4.2}$$

where $Q_i(t)$ is the unreliability of component i. From the inequality [3]

$$1 - \sum_{i=1}^{n} Q_i(t) \leq \prod_{i=1}^{n} (1-Q_i(t))$$

$$\leq 1 - \sum_{i=1}^{n} Q_i(t) + \frac{1}{2} (\sum_{i=1}^{n} Q_i(t))^2 \tag{4.3}$$

and (4.2) we conclude that for small $Q_i(t)$, $i=1,...,n$, the following approximate expression may be used

$$Q(t) \approx \sum_{i=1}^{n} Q_i(t) \tag{4.4}$$

From (1.14) and (4.1) the following expression for the system failure transition rate is deduced

$$\lambda(t) = \sum_{i=1}^{n} \lambda_i(t) \tag{4.5}$$

The expressions for the system availability and unavailability have the same general form as the expressions for system reliability and unreliability, respectively. These expressions are formally obtainable by replacing all letters R by A in (4.1) and by replacing all letter Q by U in (4.2), (4.3) and (4.4). The expressions for the steady-state values of the corresponding indices are

$$A = \prod_{i=1}^{n} A_i \tag{4.6}$$

and

$$U \approx \sum_{i=1}^{n} U_i \tag{4.7}$$

Availabilities and unavailabilities of system components in (4.6) and (4.7) are (Section 2.2)

$$A_i = \frac{\mu_i}{\lambda_i + \mu_i} \ , \quad U_i = \frac{\lambda_i}{\lambda_i + \mu_i} \ , \quad i = 1,...,n. \tag{4.8}$$

with λ_i and μ_i denoting failure transition and renewal rates of component i if both the sound state and renewal state times are exponentially distributed. If these state residence times are generally distributed, then (Section 2.4)

$$\lambda_i = \frac{1}{m_i} \ , \quad \mu_i = \frac{1}{r_i} \tag{4.9}$$

with m and r being the MTTF and MTTR of system components.

There is only one sound system state for a series connected function graph – the state in which all components are sound. The system failure transition rate equals the total rate of transition from this state to another system states. Consequently, this transition rate is

$$\lambda = \sum_{i=1}^{n} \lambda_i \tag{4.10}$$

For components with exponentially distributed state residence times (4.10) follows from (4.5).

System failure frequency may be determined by multiplying the probability of the sound system state by the system failure transition rate (refer to (3.57)).

With regard to (4.10) and (4.6) we have

$$f = \left(\sum_{i=1}^{n} \lambda_i \right) \prod_{i=1}^{n} A_i = \sum_{i=1}^{n} f_i \prod_{j} A_j \tag{4.11}$$

Index j in the last term in (4.11) applies to all system components except component i. f_i is the failure frequency of component i. In many cases in practice the availabilities of system components are very close to 1. In such cases, (4.11) may be replaced by approximate expression

$$f \approx \sum_{i=1}^{n} f_i \tag{4.12}$$

Example 4.1
The function graph is a series connection of two branches. The dependability indices of the components associated with the graph branches are given in Table 4.1. The steady-state values of system dependability indices should be calculated.

Table 4.1. Dependability indices of system components

m_1 , h	m_2 , h	r_1 , h	r_2 , h
3500	4000	10	8

We have

$$\lambda_1 = \frac{1}{m_1} = 2.857 \times 10^{-4} \text{ fl./h} \qquad \lambda_2 = \frac{1}{m_2} = 2.50 \times 10^{-4} \text{ fl./h}$$

$$\mu_1 = \frac{1}{r_1} = 0.1 \text{ ren./h} \qquad \mu_2 = \frac{1}{r_2} = 0.125 \text{ ren./h}$$

$$\lambda = \lambda_1 + \lambda_2 = 5.36 \times 10^{-4} \text{ fl./h} \qquad R(t) = \exp(-5.36 \times 10^{-4} t)$$

$$A_1 = \frac{\mu_1}{\lambda_1 + \mu_1} = 0.9971 \qquad A_2 = \frac{\mu_2}{\lambda_2 + \mu_2} = 0.9980$$

$$U_1 = 1 - A_1 = 2.9 \times 10^{-3} \qquad U_2 = 1 - A_2 = 2.0 \times 10^{-3}$$

$$f_1 = \lambda_1 A_1 = 2.85 \times 10^{-4} \text{ fl./h} \quad f_2 = \lambda_2 A_2 = 2.49 \text{ fl./h}$$

$$A = A_1 A_2 = 0.9951 \qquad U = 1 - A = 4.9 \times 10^{-3}$$

$$U \approx U_1 + U_2 = 4.9 \times 10^{-3}$$

$$f = f_1 A_1 + f_2 A_2 = 5.33 \times 10^{-4} \text{ fl./h} \qquad f \approx f_1 + f_2 = 5.34 \times 10^{-4} \text{ fl./h}$$

As can be observed, the approximate expressions for U and f have given practically accurate results. ☐

4.1.2
Parallel Connection

The only inoperable state of a system modeled by a set of n parallel connected branches is the state when all system components are simultaneously under renewal. By applying the graph terminology, the interruption of the connection between the source and sink node of the corresponding function graph will happen if all graph branches are interrupted. Consequently, the unreliability of the system, if its components are independent, equals

$$Q(t) = \prod_{i=1}^{n} Q_i(t) \tag{4.13}$$

and the reliability is

$$R(t) = 1 - \prod_{i=1}^{n} Q_i(t) \tag{4.14}$$

The system failure transition rate is obtainable from (4.14) by applying (1.13) and (1.6). There are no such simple expressions for this index as those it was possible to deduce for the series connection discussed previously.

Expressions for the system availability and unavailability are, again, obtainable from (4.14) and (4.13) by formally replacing letters R by A and Q by U. For illustration, the expressions for the steady-state system unavailability and availability are

$$U = \prod_{i=1}^{n} U_i \tag{4.15}$$

with the same meanings of symbols as previously.

$$A = 1 - \prod_{i=1}^{n} U_i \tag{4.16}$$

The system steady-state failure frequency can be calculated as the frequency of leaving the system down state which follows from the frequency balance rule (Sections 2.2 and 3.5). As the down state is abandoned by completing the renewal of any of the system components, the total system renewal rate equals the sum of the component renewal rates

$$\mu = \sum_{i=1}^{n} \mu_i \tag{4.17}$$

With regard to (4.17) and (4.15), the failure frequency equals

$$f = (\sum_{i=1}^{n} \mu_i) \prod_{i=1}^{n} U_i = \sum_{i=1}^{n} f_i \prod_{j} U_j \tag{4.18}$$

Index j applies to all system components except component I. The system MTTR equals, as follows from (4.17) and (2.33)

$$r = (\sum_{i=1}^{n} \mu_i)^{-1} \tag{4.19}$$

By comparing the expressions derived above with the expressions obtained for the series connection we conclude that these two connections are *dual*. That is, the reliability expressions of the series connection are of the same form as the expressions for the unreliability of the parallel connection. The same is the case with the expressions for availability and unavailability, as well as for system failure transition and renewal rates. This conclusion is a generalization of that already drawn for systems with two units (Section 3.6).

Example 4.2
Consider the system whose functional graph is composed of two parallel connected branches (components) with the data from Table 4.1. We should determine the dependability indices of this system.

Using (4.13) to (4.19) and component indices already calculated in the preceding example, we obtain

$$Q(t) = (1-R_1(t))(1-R_2(t)) = 1-R_1(t)-R_2(t)+R_1(t)R_2(t)$$

$$R_1(t) = \exp(-2.86\times10^{-4}t) , \qquad R_2(t) = \exp(-2.50\times10^{-4}t)$$

$$U = U_1U_2 = 5.72\times10^{-6} , \qquad f = f_1U_2 + f_2U_1 = 1.26\times10^{-6} \text{ fl./h} ,$$

$$r = \frac{r_1r_2}{r_1+r_2} = 4.44 \text{ h}$$

By comparing the results obtained with those previously determined for the series connection we can get a clear idea on the very favorable effects of the redundancy contained in the parallel connection on the reliability and availability of a system.

□

4.1.3
Series–Parallel Connections

Dependability indices of networks constructed from various combinations of series and parallel connected branches are determinable by successively applying the expressions derived for series and parallel connections when considered individually. For each such connection, taken as a single entity, the dependability indices are determined. In such a way, series and parallel branches are replaced by equivalent single branches. These are further combined in the same way until the whole network is finally reduced to a single equivalent branch between the source and sink node. The dependability indices of this branch are, then, the indices of the engineering system under consideration. For demonstration, the aforementioned approach is applied to the graph depicted in Fig. 4.2.

The equivalent branch for series connected branches 3, 4 and 5, say branch 6, has the following dependability indices

$$R_6(t) = R_3(t)R_4(t)R_5(t)$$

$$\lambda_6(t) = \sum_{i=3}^{5} \lambda_i(t)$$

$$A_6 = \prod_{i=3}^{5} A_i \qquad U_6 \approx \sum_{i=3}^{5} U_i \qquad f_6 \approx \sum_{i=3}^{5} f_i$$

The function graph is now reduced to the form displayed in Fig. 4.3 *a)*. The structure obtained is simplified by merging branches 2 and 6 into a single equivalent branch, let as say branch 7. The indices of this branch are, in accordance to the previously derived expressions for parallel connections

$$Q_7(t) = Q_2(t)Q_6(t) = Q_2(t)(1 - R_6(t))$$

$$A_7 = 1 - U_2 U_6 = 1 - U_2(1 - A_6)$$

$$f_7 = f_2 U_6 + f_6 U_2 \qquad \lambda_7 = \frac{f_7}{A_7}$$

After branch 7 is defined, the graph is converted into a series connection of two branches, as shown in Fig. 4.3 *b)*. Finally, branches 2 and 7 are merged into a single branch, say branch 8 (Fig. 4.3 *c)*), to terminate the calculation. Using the expressions deduced for series connections we obtain for branch 8

$$R_8(t) = R_1(t)R_7(t) = R_1(t)(1 - Q_7(t))$$

$$A_8 = A_1 A_7 \qquad U_8 = 1 - A_8 \qquad f_8 \approx f_2 + f_7$$

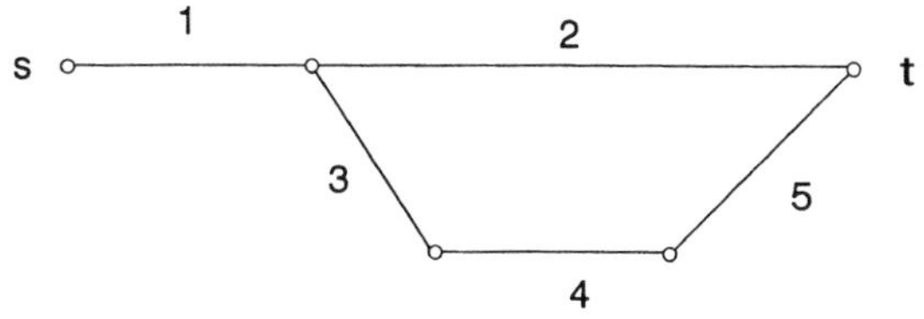

Fig. 4.2 Series-parallel function graph

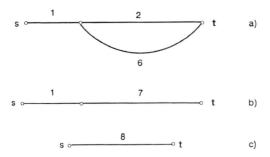

a)

b)

c)

Fig. 4.3 Steps of function graph reduction

As stated before, branch 8 indices are the dependability indices of the system being studied.

4.2
Composite Structures

Graphs not consisting of sets of series and parallel connected branches cannot be reduced to an equivalent branch by a straightforward application of the expressions derived previously. The bridge diagram displayed in Fig. 4.4 is an elementary example of such a graph.

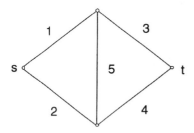

Fig. 4.4 Bridge-type graph

The calculation of the dependability indices of the aforementioned graphs can be performed only after some transformations converting them into dependability equivalent series and parallel structures.

4.2.1
General Expressions for the Steady-State Failure Frequency

Eqs. (3.54) which are valid for Markov processes as well as for two-state processes with generally distributed state residence times, as discussed in Section 2.2, can be effectively applied to the graph models. With reference to (3.60) we can write [4]

$$f = \sum_k \pi_{Ak} \lambda_k \qquad (4.20)$$

Symbol π_{Ak} denotes the probability that the system (network) *and* component (branch) k are in the sound state which transits to system failure state by failure of component (branch) k. Index k runs over all system (network) components (branches). For independent components we can write

$$\pi_{Ak} = P_{Ak} A_k \qquad (4.21)$$

P_{Ak} is the probability that the system is in the sound state *if* component k is sound and fails by failure of this component. With reference to (4.20) and (4.21) we obtain

$$f = \sum_k P_{Ak} f_k \qquad (4.22)$$

Probability P_{Ak} can be determined as

$$P_{Ak} = \Pr\{S|k\} - \Pr\{S|k'\} \qquad (4.23)$$

The first term on the r.h.s. of (4.23) is the probability that the system is sound given component k is sound. The second term is the probability that the system is sound given component k has failed. The subtraction takes into account that failure of component k does not necessarily cause the system failure. Eq. (4.23) yields the cumulative probability of these sound system states associated with the sound state of component k which transit to failure system states if component k fails.

The system (network) availability, $A = \Pr\{S\}$, is a linear combination of components availabilities and their products. From (4.23) it follows that P_{Ak} is obtainable by subtraction of system availability A determined for $A_k = 0$ from this availability determined for $A_k = 1$. If we subtract from A its value for $A_k = 0$ an expression will be obtained containing only terms including availability A_k as all terms not containing A_k are eliminated by subtraction. If we, now, set $A_k = 1$ in the constructed expression, the factor multiplying A_k in the expression for A is obtained. From the previous it follows that P_{Ak} is this factor and that it can be determined as the partial derivative of system availability with regard to availability A_k. Consequently, (4.22) can be written in the form

$$f = \sum_k \frac{\partial A}{\partial A_k} f_k \qquad (4.24)$$

The system failure frequency may alternatively be determined as the cumulative frequency of transitions from failure to sound system states which results from the frequency balance rule. The following expressions hold, analogous to those derived previously

$$f = \sum_k \pi_{Uk}\, \mu_k \tag{4.25}$$

$$f = \sum_k P_{Uk}\, f_k \tag{4.26}$$

$$P_{Uk} = \pi_{Uk}\, U_k \tag{4.27}$$

$$P_{Uk} = \Pr\{S'|k'\} - \Pr\{S'|k\} \tag{4.28}$$

$$f = \sum_k \frac{\partial U}{\partial U_k}\, f_k \tag{4.29}$$

Here, π_{Uk} is the probability that the system *and* component k are in failure states and that the system is restorable by renewal of component k. P_{Uk} is the probability that the system is in a failure state *given that* component k has failed and that the system is restorable by renewal of component k.

Example 4.3
Expression (4.11) is simply obtainable from (4.6) by applying (4.24). Also, (4.17) is derived from (4.15) if (4.29) is used.

□

4.2.2
Application of the Total Probability Theorem

According to the *total probability theorem* the probability of an event, say e, composed of a set of mutually exclusive events e_i can be determined as

$$\Pr\{e\} = \sum_i \Pr\{e|e_i\}\, \Pr\{e_i\} \tag{4.30}$$

where i applies to all component events. It is presumed that the probabilities $\Pr\{e_i\}$ add to 1.

By utilizing (4.30), the reliability of the graph in Fig. 4.4 may be calculated by means of the following expression

$$R(t) = R_{S5}(t)R_5(t) + R_{\bar{S}5}(t)(1 - R_5(t)) \tag{4.31}$$

with $R_{S5}(t)$ denoting the reliability of the graph when branch 5 is sound and $R_{\bar{S}5}(t)$ being the graph reliability given branch 5 is in failure state. The aforementioned reliabilities are simply determinable from the graphs obtained by open removal and short cut of branch 5 in Fig. 4.4. These graphs are presented in Fig. 4.5 *a)* and *b)*, respectively. They consist of only series and parallel connected branches and can be simply analyzed for any dependability index using the approach discussed in the previous section.

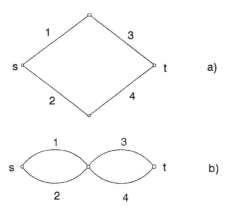

Fig. 4.5 Graphs for calculating $R_{S5}(t)$ (graph *a)*) and $R_{S5}(t)$ (graph *b)*)

The system availability equals, with regard to (4.30) and (4.31)

$$A(t) = A_{S5}(t)A_5(t) + A_{S5\prime}(t)(1-A_5(t)) \tag{4.32}$$

with $A_{S5}(t)$ and $A_{S5\prime}(t)$ being the availabilities calculated for graphs in Fig. 4.5 *b)* and *a)*.

The failure frequency of the system can be determined by applying (4.22) or (4.26). For illustration, P_{A5} in (4.23) equals

$$P_{A5} = A_{S5} - A_{S5\prime} \tag{4.33}$$

The terms on the r.h.s. of (4.33) are the steady-state availabilities for graphs in Fig. 4.5 *b)* and *a)*.

By a successive application of the approach based upon the total probability theorem, arbitrarily complex graphs can be handled. However, there are other methods for such an analysis based upon relatively simple and general, easily programmable algorithms which are therefore preferred. Some of them will be presented in the forthcoming sections.

4.2.3
Minimal Paths

The *minimal path* is a series connection of graph branches connecting the source and sink node in which no node is traversed more than once. For illustration, consider the bridge-type graph depicted in Fig. 4.6. The walk *gccd* is not a minimal path as it traverses node 2 twice. However, *gce* is a minimal path as nodes 2 and 3 belonging to this path are traversed only once. The *cardinality* of a minimal path is defined as the number of branches contained in the path. The highest cardinality a minimal path for a certain network may have equals the number of network nodes minus 1. Clearly, such a case arises if all network branches are series connected to build a single minimal path.

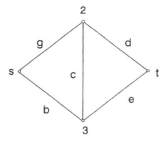

Fig. 4.6 Graph of a bridge-type network

There are various methods for determining minimal paths [11,12]. A simple way to deduce all minimal paths of a network is via the *connection matrix*. Consider a network having n nodes. The connection matrix is a square n by n matrix having zero elements along its main diagonal. Element e_{ij} being in row i and column j of this matrix is the branch connecting nodes i and j. The source node is numbered as the first, and the sink node as the last of network nodes. If there are several branches connecting nodes i and j, they all appear as the element in row i and column j of the connection matrix but separated by the sign + which indicates the word *or*. For bidirectional networks whose branches may be walked over in both directions, the connection matrix is symmetrical.

For demonstration, the connection matrix for the network with the graph in Fig. 4.6 is

$$[C] = \begin{bmatrix} 0 & g & b & 0 \\ g & 0 & c & d \\ b & c & 0 & e \\ 0 & d & e & 0 \end{bmatrix} \qquad (4.34)$$

It was presumed that all branches are bidirectional. The property of the connection matrix is that its kth power yields the paths of cardinality k among network nodes. The second and third power of matrix [C] for the network under consideration are

$$[C]^2 = \begin{bmatrix} 0 & bc & gc & dg+be \\ bc & 0 & bg+de & ce \\ gc & bg+de & 0 & cd \\ dg+be & ce & cd & 0 \end{bmatrix} \quad (4.35)$$

$$[C]^3 = \begin{bmatrix} 0 & bde & deg & ceg+bcd \\ bde & 0 & 0 & beg \\ deg & 0 & 0 & bdg \\ ceg+bcd & beg & bdg & 0 \end{bmatrix}. \quad (4.36)$$

The multiplication of matrices has to obey a specific rule. If a branch appears more than once in a product this product should be omitted. It indicates a nonminimal path as a branch and its nodes are traversed more than once. Each product indicates a minimal path and the branches that are included in this path. The symbol of multiplication should stand for *and*.

In the example being discussed the maximum cardinality of minimal paths does not exceed 3. Hence, the higher powers of [C] will indicate only nonminimal paths. As can be concluded, there are two minimal paths of cardinality 2 (*dg* and *be*) and two minimal paths of cardinality 3 (*ceg* and *bcd*) connecting the source and sink node. The reader can visually verify this result by inspecting the graph in Fig. 4.6.

If we are interested only in the minimal paths connecting the source node with other nodes and, particularly, with the sink node, which is the case in the dependability analysis, the powers of the connection matrix contain some surplus information. The elements of the first row of the connection matrix powers that yield these paths can be deduced with significantly less computational effort if the following approach is used [7]: the first row of $[C]^2$ is obtained by multiplying from the right the first row of [C] by [C]. The first row of $[C]^3$ is obtained by multiplying from the right the first row of $[C]^2$, determined in the preceding step, by [C], etc.

The connection between the source and sink node will be interrupted if all minimal paths are interrupted. From the aforesaid we conclude that the graph in Fig. 4.6 may, from the dependability standpoint, be replaced by the parallel connection of minimal paths shown in Fig. 4.7.

As noted, minimal paths contain common branches which means that minimal paths are not independent. Consequently, the approach applied in Section 4.1 to independent series and parallel connected branches cannot be used but methods for encompassing a more complex correlation among graph branches and paths must be used.

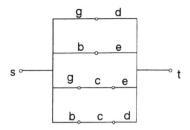

Fig. 4.7 Equivalent minimal paths graph

Consider a set of events e_i, $i=1,...,m$, which all may be dependent, generally. The probability that at least one of the events e_i has happened equals

$$\Pr\{e\} = \sum_{i=1}^{m} \Pr\{e_i\} - \sum_{i=1}^{m-1}\sum_{j=i+1}^{m} \Pr\{e_i e_j\} +$$

(4.37)

$$\sum_{i=1}^{m-2}\sum_{j=i+1}^{m-1}\sum_{k=j+1}^{m} \Pr\{e_i e_j e_k\}...+ (-1)^{m-1}\Pr\{\prod_{i=1}^{m} e_i\}$$

This is the well known *inclusion exclusion formula* for complex events. The first term in (4.37) yields the upper bound of $\Pr\{e\}$, while the first two terms provide the lower bound. The first three terms in (4.37) yield a better upper bound and the first four terms yield a better lower bound. Still better bounds are provided by the first five and six terms, etc.

Let e_i be the reliability of minimal path i. Event e is, then, by definition, the reliability of the minimal path graph, i.e., of the system which is modeled by this graph. The probabilities figuring in (4.37) are then

$$\Pr\{e_i\} = R_{pi}(t) = \prod_{j} R_{ij}(t)$$

(4.38)

where $R_{pi}(t)$ designates the reliability of minimal path i and $R_{ij}(t)$ are the reliabilities of the branches belonging to path i. Index j applies to all these branches. The probability of several minimal paths being simultaneously reliable is determined by taking into account that minimal paths might have common branches. For example, the probability that paths i, j and k are simultaneously sound until time t equals

$$\Pr\{e_i e_j e_k\} = \prod_{s} R_s(t)$$

(4.39)

where index s applies to all branches contained in paths i, j and k, but each of them taken into account only once. For demonstration, the probability of all four minimal paths in Fig. 4.7 being reliable at the same time equals

$$\text{Pr}\{ \prod_{i=1}^{4} e_i \} = R_g(t)R_b(t)R_c(t)R_d(t)R_e(t) \tag{4.40}$$

The system availability is determinable using (4.37) for e_i being the event that minimal path i is available at time t. Hence,

$$\text{Pr}\{e_i\} = A_{pi}(t) = \prod_j A_{ij}(t) \tag{4.41}$$

where $A_{pi}(t)$ denotes the availability of minimal path i while $A_{ij}(t)$ is the availability of branch j of path i. The expression for the probability of minimal paths i, j and k all being available equals, in analogy to (4.39)

$$\text{Pr}\{e_i e_j e_k\} = \prod_s A_s(t) \tag{4.42}$$

with s covering all branches belonging to minimal paths under consideration, but each branch taken into account only once.

The steady-state failure frequency of the system when modeled by minimal paths may be calculated by applying (4.22) and (4.23) to the corresponding system minimal path graph [6]. Probability $\text{Pr}\{S|k\}$ equals the steady-state availability value of the graph obtained by the short cut of removing branch k of the system graph. Probability $\text{Pr}\{S|k'\}$ is obtained as the steady-state availability value of the graph constructed from the system graph by open removal of branch k. For illustration, probabilities $\text{Pr}\{S|g\}$ and $\text{Pr}\{S|g'\}$ for the graph in Fig. 4.7 are determined as steady-state availabilities for graphs in Fig.4.8 $a)$ and $b)$, respectively.

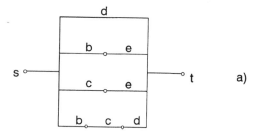

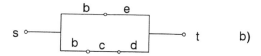

Fig. 4.8 Graphs for deducing $\text{Pr}\{S|g\}$ (graph $a)$) and $\text{Pr}\{S|g'\}$ (graph $b)$)

Example 4.4

Consider the graph in Fig. 4.7. Expressions for the terms in (4.37) for the steady state availability are presented for demonstration. With regard to (4.41) and (4.42) we have

$$\Pr\{e_1\} = A_{p1} = A_g A_d , \qquad \Pr\{e_2\} = A_{p2} = A_b A_e .$$

$$\Pr\{e_3\} = A_{p3} = A_g A_c A_e , \qquad \Pr\{e_4\} = A_{p4} = A_b A_c A_d ,$$

$$\Pr\{e_1 e_2\} = A_g A_d A_b A_e , \qquad \Pr\{e_1 e_3\} = A_g A_d A_c A_e$$

$$\Pr\{e_1 e_4\} = A_g A_d A_b A_c , \qquad \Pr\{e_2 e_3\} = A_b A_e A_g A_c$$

$$\Pr\{e_2 e_4\} = A_b A_e A_c A_d , \qquad \Pr\{e_3 e_4\} = A_g A_c A_e A_b A_d ,$$

$$\Pr\{e_1 e_2 e_3\} = \Pr\{e_1 e_2 e_4\} = \Pr\{e_2 e_3 e_4\} = \Pr\{e_1 e_2 e_3 e_4\} = \Pr\{e_3 e_4\}$$

$$\Pr\{e_1 e_2 e_3\} = \Pr\{e_1 e_2 e_4\} = \Pr\{e_2 e_3 e_4\} = \Pr\{e_1 e_2 e_3 e_4\} = \Pr\{e_3 e_4\}$$

$\square$

4.2.4
Unreliable Nodes

If a node, say node k, is exposed to failures, all minimal paths traversing this node are interrupted by each failure of node k. Therefore, for a proper assessment of the effects of such a failure, all minimal paths containing any branch incident to node k

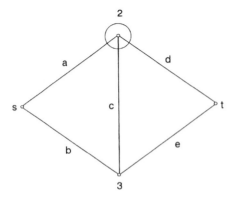

Fig. 4.9 Graph with an unreliable node

should be complemented by a fictitious branch characterized by node k dependability indices [8]. This branch interrupts the associated minimal paths at each node k failure. Consider for illustration the graph in Fig. 4.9 having unreliable node 2. Minimal paths of the same graph but with reliable nodes are *ad, ace, be,* and *bcd.* Minimal paths through node 2 are all paths containing any of branches *a, c,* and *d.* Thus, with regard to the aforesaid, the minimal paths of the graph with unreliable node 2 are: *ad2, ace2, be,* and *bcd2* with symbol 2 denoting the fictitious branch reflecting node 2 dependability impacts.

4.2.5
Multiple Sources and Sinks

Function graphs are often used to model actual engineering systems containing multiple sources and consumers which are interconnected through a network. The question is how to analyze the dependability of such a system by applying the function graph approach. The answer is relatively simple. All sources are modeled as graph branches connecting the associated network nodes to a single source node. Consider the function graph depicted in Fig. 4.10. Let us assume that system sources are connected to the network at nodes 2 and 3 while nodes 4 and 5 are consumer nodes. Branches *a* and *b* model the sources and their dependability indices and *s* is the graph source node. Using the graph presented, the dependability of the system in performing various tasks can be determined. If the dependability of the supply of the consumer at node 4 is of the prime concern, the minimal paths leading from the source node to node 4 are to be determined as described before, to obtain the desired information. The same is the case with the consumer at node 5 if this consumer is of interest. As shown in Section 4.2, the sets of minimal paths to nodes 4 and 5 are simultaneously obtainable by applying the connection matrix approach. These paths for node 4 are *ad, aceg, bcd, beg.* Minimal paths for node 5 are *adg, ace, bcdg, be.*

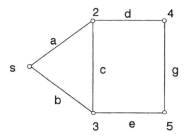

Fig. 4.10 System with multiple sources and sinks

4.2.6
Minimal Cuts

The graph *cut* is a set of branches whose simultaneous open removal interrupts the connection between the source and sink graph nodes. The *minimal cut* is the cut which does not contain any other cut. The cut *order* equals the number of branches it is built from. The highest order a cut can have equals the difference between the number of graph branches and nodes, increased for 2. The parallel branches are herein counted as a single branch. Two cuts are said to be dependent if they contain any common branch.

There are various methods for deducing minimal cuts [10,11,13]. We shall describe the method based upon the enumeration of minimal paths [10].

For a graph represented by the set of minimal paths a *minimal path matrix* can be defined. This is an *m* by *n* matrix where *m* is the number of minimal paths and *n* is the number of graph branches. Element v_{ji} of this matrix equals 1 if branch *i* is included in minimal path *j* and is zero otherwise. For illustration, the minimal path matrix for the graph in Fig. 4.7 is, if the branches are ordered following the sequence *g, b, c, d, e,*

$$[P] = \begin{bmatrix} 1 & 0 & 0 & 1 & 0 \\ 0 & 1 & 0 & 0 & 1 \\ 1 & 0 & 1 & 0 & 1 \\ 0 & 1 & 1 & 1 & 0 \end{bmatrix} \qquad (4.43)$$

Each column of [P] indicates the paths to which the corresponding branch belongs. If the elements of a column are all units, the corresponding branch figures in all paths and, thus, the failure of this branch interrupts all minimal paths and the connection between the source and sink node. Consequently, such a branch is a minimal cut of the first order. The higher order cuts are determined by addition of columns of [P] by applying Boolean algebra. As is known, the rules for addition in Boolean algebra are

$$\begin{aligned} 1 + 1 = 1 \qquad & 1 + 0 = 1 \\ & \qquad\qquad\qquad (4.44) \\ 0 + 1 = 1 \qquad & 0 + 0 = 0 \end{aligned}$$

By adding one column to another, a column vector is obtained indicating the paths including any or both of the branches associated with the columns. If the resulting column vector elements are all unity, these two branches build a second order cut, as if they are both down the source and sink node are separated from one another. If neither of these branches is a first order cut, the deduced cut is a minimal cut. The approach described is applied to the matrix [P] using a sequential procedure enabling the identification of all minimal cuts. The procedure is the following:

a). By inspection of [P], the columns having all unit elements are identified. The branches associated with these columns are first order cuts. The aforementioned columns are omitted from [P] as they, since they correspond to the first order minimal cuts, cannot build higher order minimal cuts and need not be combined with the remaining columns of [P]. Let us denote by [P'] the reduced minimal path matrix.

b). The first column of [P'] is added obeying (4.44) to the remaining matrix columns one by one to deduce the second order minimal paths generated by the first branch in [P']. The same procedure is repeated with the second, third,..., column, which are added to succeeding columns. By following this procedure all second order minimal cuts are determined.

c). Pairs of columns of [P'] are successively added to the succeeding columns to deduce the third order cuts. The cuts containing the second order minimal cuts deduced in step b) are omitted. The remaining cuts are the minimal third order cuts.

d). Triples of columns of [P'] are added to the succeeding columns to determine fourth order cuts. We extract again the cuts that do not contain any lower order minimal cut determined in previous steps. These cuts are the fourth order minimal paths.

e). The procedure described above is followed until the highest order minimal cuts are deduced.

In many cases in practice the minimal cuts up to the third order are determined as the probability of the overlapping of more than three failure states is extremely low and such events may be discarded.

For demonstration, minimal cuts of the graph in Fig. 4.6, deduced by applying the previously described approach to the minimal paths graph in Fig. 4.7, are

$$gb \qquad de \qquad gce \qquad bcd \tag{4.45}$$

We leave it to the reader to verify this result.

The connection between the source and the sink node is interrupted if all branches belonging to any minimal cut are interrupted. Thus, the *minimal cut function graph* is built by series connection of minimal cuts which are represented as corresponding sets of parallel connected branches. For illustration, Fig. 4.11 depicts the minimal cut based graph equivalent to the graphs in Figs. 4.6 and 4.7.

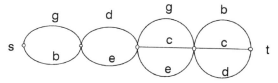

Fig. 4.11 Minimal cut graph equivalent to the graphs in Fig.4.6 and Fig.4.7

Minimal cuts may contain common branches with other minimal cuts, as may be observed from (4.45). Therefore, they are dependent, generally. Owing to this fact, the minimal cut graph cannot be reduced to a single equivalent branch by applying the technique described in Section 4.1. The dependability indices have to be calculated by applying the inclusion exclusion formula (4.37) in a similar way as for the minimal paths.

Let event e_i be the failure of minimal cut i before time t. Then

$$\Pr\{e_i\} = Q_{ci}(t) = \prod_j Q_{ij}(t) \tag{4.46}$$

with $Q_{ci}(t)$ denoting the unreliability of minimal cut i and $Q_{ij}(t)$ being the unreliability of branch j belonging to cut i. Index j runs over all branches building cut i. The probability of several minimal cuts being down simultaneously, say cuts i, j, k, equals

$$\Pr\{e_i e_j e_k\} = \prod_s Q_s(t) \tag{4.47}$$

where index s applies to all branches contained in cuts i, j, k, each branch being taken into account only once.

The expression for the graph unavailability is obtainable from this expression for unreliability by replacing symbol Q by U with the same meaning of indices as before. The expression for the steady-state unavailability values is obtained from the expression for the time-specific unavailability by formally replacing the symbols for time-specific unavailabilities with the symbols for their steady-state values. If the unavailabilities of branches are very low, which is often the case for many engineering systems, the expression

$$U \approx \sum_{i=1}^{m} U_{ci} \tag{4.48}$$

provides a good upper bound. In (4.48) U_{ci} is the unavailability of minimal cut i and m is the total number of minimal cuts.

The system failure frequency may be calculated using (4.26) and (4.28). The probabilities in (4.28) can simply be determined [6] from the graph in Fig. 4.11. Probability $\Pr\{S \mid k\}$ equals the steady-state unavailability of this graph with branch k missing. Probability $\Pr\{S \mid k\}$ equals the steady-state unavailability of the graph obtained from the graph in Fig. 4.11 by omission of all minimal paths containing branch k for if k is up all cuts containing k are up. For illustration, Fig. 4.12 displays the graphs used to determine $\Pr\{S \mid g\}$ and $\Pr\{S \mid g\}$ for the graph in Fig. 4.11.

System failure frequency can be determined based upon minimal cut representation by utilizing the more condensed expression [5,9]

$$f = \sum_{i=1}^{m} U_{ci}\, \mu_{ci} - \sum_{i=1}^{m-1} \sum_{j=i+1}^{m} U_{cij}\, \mu_{cij} \ldots + $$
$$+ (-1)^{m-1} U_{cij\ldots m}\, \mu_{cij\ldots m} \tag{4.49}$$

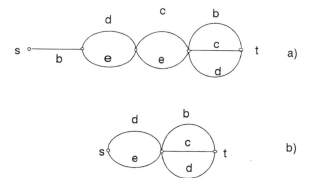

Fig. 4.12 Graphs for calculating: a) $\Pr\{S'|g'\}$ and b) $\Pr\{S'|g\}$

$U_{cij...}$ is the probability of the simultaneous failure state of cuts $i, j, ...$. This probability is calculated by multiplying the unavailabilities of all branches contained in the corresponding minimal cuts, each branch being taken only once. Parameter $\mu_{ij...}$ equals the sum of renewal transition rates of all branches belonging to cuts $i, j, ...$, each branch being taken into account only once. The first term in (4.49) provides a good upper bound for the system failure frequency in most cases in practice. The first two terms yield the lower bound for this frequency.

Example 4.5
The terms in (4.37) for the steady-state unavailability of the graph in Fig. 4.6 should be written in general form, as illustration.
 With regard to (4.46) and (4.47) we have

$$\Pr\{e_1\} = U_{c1} = U_g U_b , \qquad \Pr\{e_2\} = U_{c2} = U_d U_e ,$$

$$\Pr\{e_3\} = U_{c3} = U_g U_c U_e , \qquad \Pr\{e_4\} = U_{c4} = U_b U_c U_d ,$$

$$\Pr\{e_1 e_2\} = U_{c12} = U_g U_b U_d U_e , \qquad \Pr\{e_1 e_3\} = U_{c13} = U_g U_b U_c U_e ,$$

$$\Pr\{e_1 e_4\} = U_{c14} = U_g U_b U_c U_d , \qquad \Pr\{e_2 e_3\} = U_{c23} = U_d U_e U_g U_c ,$$

$$\Pr\{e_2 e_4\} = U_{c24} = U_d U_e U_b U_c , \qquad \Pr\{e_3 e_4\} = U_{c34} = U_g U_c U_e U_b U_d ,$$

$$U_{c123} = U_{c124} = U_{c234} = U_{c1234} = U_{c34}$$

□

4.2.7
Network Partition and Equivalents

To lighten the computational burden in determining minimal paths it may be helpful in some cases to separate the network into subnetworks. The network is decomposed into two subnetworks, say A and B, through a minimal cut. Subnetwork A contains the branches of this cut and branches that are on the same side as the source node. Subnetwork B contains the remaining branches. The nodes at the ends of the minimal cut branches through which subnetworks A and B are interconnected build the set of common nodes. The general idea is to calculate separately the minimal paths for subnetworks leading from source and sink node to the common nodes and then to combine them to obtain the minimal paths of the network as a whole. For illustration, consider a modification of the ARPA telecommunications network depicted in Fig. 4.13 [14]. Fig. 4.14 presents the decomposition of the aforementioned network through cut *edj* yielding two common nodes: 3 and 7. In Fig. 4.15 the subnetworks are modeled by minimal paths connecting the source and sink node to the common nodes. These paths are determined for both subnetworks separately. The network minimal paths are obtainable by series connecting each minimal path of subnetwork A leading to a specific common node to each minimal path of subnetwork B leading to the same common node. There are, in total, 25 minimal paths for the example being examined. 16 minimal paths traverse through node 3, and 9 through node 7. All these paths are minimal as neither network node is walked over more than once.

By partitioning, the network is separated into two less composite networks which are handled separately. That might be beneficial with regard to the consumption of computer resources. There is another advantage of network partition which may be used in practical applications. Namely, if various modifications of a network are analyzed associated with one subnetwork exclusively, then the other subnetwork remains intact and its minimal path model may be used as a firm network equivalent in all these analyses.

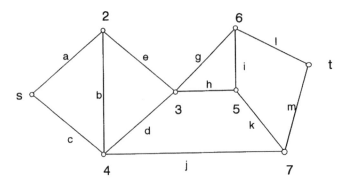

Fig. 4.13 Modified ARPA network

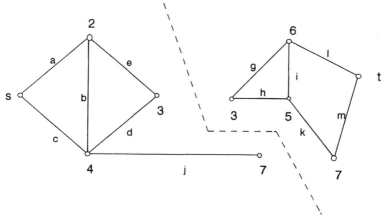

Fig. 4.14 Partition through cut *edj*

However, if a network is partitioned in such a way as to have more than two common nodes, the simple combination of subnetwork minimal paths might fail to enumerate all minimal paths. Consider, for example, the partition of the ARPA network through cut *cbdhg* generating three common nodes: 4, 5, 6 (Fig. 4.16). Minimal paths *abjkhgl* and *cjkhgl* would be overlooked by solely combining subnetwork minimal paths leading to common nodes, as in the previous case. These two paths cross twice the border separating the subnetworks. They are composed of a string of branches of subnetwork *A* leading from the source node to common node 7, a string of branches connecting common nodes 7 and 5 through subnetwork *B*, a string of branches connecting nodes 5 and 6 through subnetwork *A* and, finally, of branch *l* of subnetwork *B*, connecting node 6 to the sink node. Such minimal paths multiply traversing the border between the subnetworks and consisting of an alternating sequence of strings through subnetworks, do not allow the composition of simple equivalent schemes for subnetworks. This matter is discussed in more detail in [14].

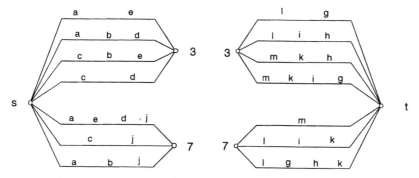

Fig. 4.15 Minimal path subnetwork equivalent schemes

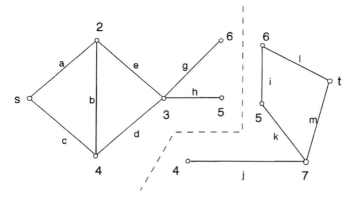

Fig. 4.16 The partition through cut *cbdhg*

4.2.8
Networks with Dependent Branches

Many engineering systems are successfully modeled by networks whose dependability performances are relatively simply analyzed by applying the minimal cut approach. However, to reflect the actual circumstances which might arise in reality, it is necessary to account for the dependence of some groups of network branches. Such circumstances include: common-cause failures from external origin, faults induced from neighboring branches, operation of protective and control devices, restricted repair, cold standby operation. Some of these circumstances have already been discussed in Section 3.6. We shall demonstrate a method enabling us to assess the dependability of branches in the minimal cut based network dependability analysis. The method uses the Kolmogorov equations to model the mutual dependence of states of network branches. This implies that exponentially distributed state residence times of branches are presumed. As discussed before, this is a restrictive presumption for time-specific network dependability indices only. The steady-state results are also valid for most nonexponential systems.

The following expression for network failure probability in terms of minimal cuts holds, based upon the formula for the union of events

$$
P = \sum_{j=1}^{n} \Pr\{ \prod_{i \in C_j} e_i \} - \sum_{j=1}^{n-1} \sum_{k=j+1}^{n} \Pr \{ \prod_{i \in C_j \cup C_k} e_i \} +
$$

$$
\dots + (-1)^{n-1} \Pr \{ \prod_{i=1}^{m} e_i \}
$$

(4.50)

In (4.50) e_i denotes the event: branch i is in nonoperating state. C_k is the set of indices of branches contained in minimal cut k. Let B be the set of indices of mutually dependent branches. Then the terms in (4.50) can be written in the form

$$\Pr\left\{\prod_{i\in(C_j\cup\ldots)} e_i\right\} = \Pr\left\{\prod_{i\in B\cap(C_j\cup\ldots)} e_i\right\}\Pr\left\{\prod_{i\in(C_j\cup\ldots)\backslash B} e_i\right\} \tag{4.51}$$

The first term on the r.h.s. of (4.51) is obtainable from the corresponding Kolmogorov equations for the Markov process describing the behavior of branches in B

$$\Pr\left\{\prod_{i\in B\cap(C_j\cup\ldots)} e_i\right\} = \sum_j p_j \tag{4.52}$$

where p_j are the probabilities of Markov process states that include event $\prod e_i$, $i\in B\cap(C_j\cup\ldots)$.

Expressions (4.50) through (4.52) are valid for both time-specific and steady-state analysis.

As discussed in Section 3.4, the steady-state failure frequency of a system may be determined as the frequency of encountering network operating states by transitions from network nonoperating states

$$f = \sum_k\sum_i z_{ki}\, P_{ki} \tag{4.53}$$

In (4.53) z_{ki} is the transition rate of branch k nonoperating state of mode i to an operating state. P_{ki} is the probability that the network is in a nonoperating state while branch k is in a nonoperating state and that network operation is restorable by bringing branch k back to an operating state. Index k is over all network branches and i applies to all nonoperating modes of branch k.

By analogy to (4.28)

$$P_{ki} = P_{eki} - P_{cek} \tag{4.54}$$

P_{eki} is the probability that the network is in a nonoperating state while branch k is in a nonoperating state of mode i. P_{cek} is the probability that the network is in a nonoperating state while branch k is in an operating state. P_{eki} is obtainable from (4.50) by substituting e_{ki} for e_k. By definition, probability P_{cek} can be deduced by applying (4.50) to minimal cuts not containing branch k. After performing the subtraction in (4.54), the terms which are not associated with branch k will disappear. Hence, the expression obtained is of the same general form as (4.50) but it contains only the terms associated with branch k, with e_{ki} substituted for e_k.

Example 4.6

The circumstances which could arise if failure of some units leads to switching out of other units in order to protect the system from further damage are displayed in Fig. 4.17. Such *"active"* failures are typical for electrical power substations where faults

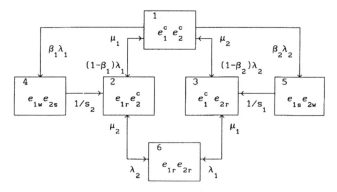

Fig. 4.17 State-transition diagram modeling active failures

causing high short circuit currents are immediately switched out by opening adjacent circuit breakers. β_k, $k=1$, 2, is the probability that the failure of element k is active. Indices w, s and r have the following meanings, respectively: waiting for repair, good but switched out, under repair. Event e_i^c is the complement to event e_i. The activated breakers are reclosed after a relatively short time s_k, $k = 1,2$, needed to isolate or remove the faulted element. This is modeled by transitions from states 4 and 5 to states 2 and 3, respectively.

The approach suggested in this paper will be illustrated using the simple network displayed in Fig. 4.18, for detailed presentation. It is presumed that branches 1 and 2 are dependent as they are prone to active failures. Further on, p_k, $k = 2,...,6$, are the probabilities of states in Fig. 4.17.

By applying (4.50) and (4.52) to the network in Fig. 4.18 we obtain

$$P = \Pr\{e_1 e_2\} + \Pr\{e_1\}\, \Pr\{e_3\} - \Pr\{e_1 e_2\}\, \Pr\{e_3\}$$

With reference to (4.52) and Fig. 4.17

$$\Pr\{e_1 e_2\} = p_4 + p_5 + p_6$$

$$\Pr\{e_1\} = p_2 + p_4 + p_5 + p_6$$

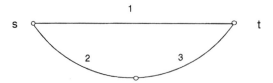

Fig. 4.18 Network sample under consideration

From the two preceding expressions it follows that

$$P = p_4 + p_5 + p_6 + p_2 \Pr\{e_3\}$$

According to (4.53) we have

$$f = \mu_1 P_{1r} + \frac{1}{s_1} P_{1s} + \mu_2 P_{2r} + \frac{1}{s_2} P_{2s} + \mu_3 P_3$$

The probabilities appearing in this expression are determinable using (4.54). For demonstration, a detailed derivation is given as follows.

As $P_{cel} = 0$, we have

$$P_{1r} = \Pr\{e_1, e_2\} + \Pr\{e_{1r}\}\Pr\{e_3\} - \Pr\{e_1, e_2\}\Pr\{e_3\} =$$

$$p_6 + (p_2 + p_6)\Pr\{e_3\} - p_6\Pr\{e_3\} = p_6 + p_2\Pr\{e_3\}$$

$$P_{1s} = \Pr\{e_1, e_2\} + \Pr\{e_{1s}\}\Pr\{e_3\} - \Pr\{e_1, e_2\}\Pr\{e_3\} =$$

$$p_5 + p_5\Pr\{e_3\} - p_5\Pr\{e_3\} = p_5$$

It should be noted that state 5 is the only state with branch 1 under switching-in-service action, which explains the previous equation.

For P_{2r} and P_{2s} the second term in the general expression for P should be discarded as it not associated with branch 2. Hence,

$$P_{2r} = \Pr\{e_1 e_{2r}\} - \Pr\{e_1 e_{2r}\}\Pr\{e_3\} =$$

$$\Pr\{e_1 e_{2r}\}(1 - \Pr\{e_3\}) = p_6(1 - \Pr\{e_3\})$$

$$P_{2s} = \Pr\{e_1 e_{2s}\} - \Pr\{e_{2s}\}\Pr\{e_3\} =$$

$$p_4 - p_4\Pr\{e_3\} = p_4(1 - \Pr\{e_3\})$$

State 4 is the only state with branch 2 under switching-in-service action. This explains the latter expression.

For P_3 the first term in the general expression for P has to be omitted

$$P_3 = \Pr\{e_1\}\Pr\{e_3\} - \Pr\{e_1 e_2\}\Pr\{e_3\} =$$

$$(p_2 + p_4 + p_5 + p_6)\Pr\{e_3\} - (p_4 + p_5 + p_6)\Pr\{e_3\} = p_2\Pr\{e_3\}$$

The expressions derived for failure probability and steady state frequency are the same as those obtainable by applying the Kolmogorov equations to the sample network. We leave it to the reader to check this.

$\square$

4.2.9
Limited Transfer Capacity Networks

In many engineering systems function graph branches have a limited transfer capacity. This is commonly the case when actual engineering networks used to deliver some kind of goods or information from source to consumer (receiver) terminals are modeled . For such networks the transfer capability in various possible failure states should be determined to establish to what extent its function is degraded in these states. The dependability indices of limited capacity networks are determinable by screening network states using enumeration or simulation techniques which will be presented later in this book. We shall present here a relatively simple method for determining the transfer capability of networks using network minimal paths. There are several efficient methods for the calculation of flows through a network proposed so far [15,16]. Our choice was motivated by the fact that the subject analysis is performed within the scope of dependability evaluation and that, therefore, the sets of minimal paths to network nodes are already available.

The method loads the branches of minimal paths to a certain terminal node until its load is covered. The transfer capacity of each path equals that of the smallest capacity branch belonging to the path. The paths are loaded following the descending reliability (availability) or ascending cardinality order, for example. The terminal loads are supplied by obeying a priority list, if any, or in any other preadopted order. The above order rules affect the supply of consumers only if their total demand exceeds the network transfer capacity. In loading various paths, branches might be traversed in alternate directions. The flows through branches are summed by regarding these directions. To demonstrate the method more clearly, we consider the bridge-type network in Fig. 4.19 with annotated branch capacities and node demands. Network branches are directed from lower to higher end node numbers. It is presumed that they have the same capacity in both directions.

The nodes are supplied in descending order of demands. The steps in network flow calculation are presented in Table 4.2.

Node No. 4 is supplied first. Path ad is used whose capacity is 2 units owing to branch d. Table 4.2 indicates that 2 capacity units of branches a and d are used. Thus, the available capacity of branch a lies within the interval $(-5, 1)$ which means that it can be loaded for 1 unit in the positive direction and for up to 5 units in the negative direction. Similarly, the capacity of branch d left for further flow transfer is within the interval $(-4, 0)$. Path ad supplies the consumer at node No. 4 with 2 units, as indicated in the last column. The required extra 1 unit flow is transferred via path be loading branches b and e with 1 flow unit. This affects the available capacity

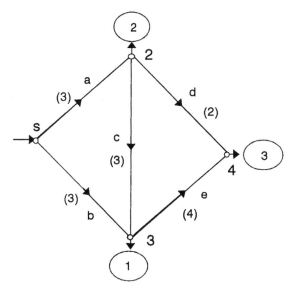

Fig. 4.19 Sample limited capacity network

of these two branches. The available capacities of the remaining branches remain unchanged. Node No. 2 is supplied via paths a and bc and node No. 3 via path b, with flows and available capacities determined using the same procedure as explained for Node No. 4 and presented in Table 4.2.

Table 4.2 . Determination of network flows

Node paths	branch:	a	b	c	d	e	load suppl.
No.4 ad	capac. used left	2 (−5,1)	0 (−3,3)	0 (−3,3)	2 (−4,0)	0 (−4,4)	2
be	used left	0 (−5,1)	1 (−4,2)	0 (−3,3)	0 (−4,0)	1 (−5,3)	1
No. 2 a	capac. used left	1 (−6,0)	0 (−4,2)	0 (−3,3)	0 (−4,0)	0 (−5,3)	1
bc	used left	0 (−6,0)	1 (−5,1)	-1 (−2,4)	0 (−4,0)	0 (−5,3)	1
No. 3 b	capac. used left	0 (−6,0)	1 (−6,0)	0 (−2,4)	0 (−4,0)	0 (−5,3)	1

The flows determined through network branches are displayed in Fig. 4.20. They are obtained by summing the flows associated with network consumer nodes.

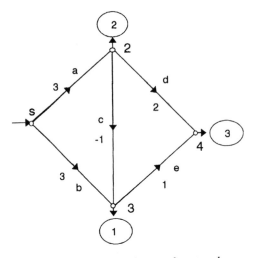

Fig. 4.20 Flows through the sample network

Problems

1. A system is composed of four identical units. Construct the function graph for this system in the following cases:
   a) System is up if at least two units are up
   b) System is up if at least three units are up

2. Determine the steady-state unavailability of the system with the graph in Fig. E4.1 using the total probability theorem

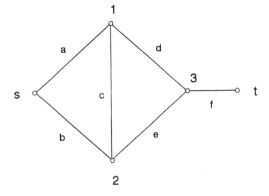

Fig. E41 Sample function graph

3. Deduce minimal paths and minimal cuts of the sample system graph in Fig. E4.1.

4. Assume that node No. 1 of the graph in Fig. E4.1 is unreliable. Determine the minimal paths and cuts of the graph. (Minimal paths are deduced first and, then, by means of them, the minimal cuts.)

5. Separate the graph in Fig. E4.1 through minimal cut *ab*. Determine minimal paths for both subnetworks to common nodes. By combining these minimal paths deduce the minimal paths of the network as a whole.

6. Suppose that the transfer capacity of branches *a,b,c,* of the network in Fig. E4.1 is 2 units each and that of branches *d,e,f* one unit each. All branches are bidirectional with the exception of branch *c* which does not allow flows from node 1 to node 2. Determine the maximum possible flow from the source node to: a) node 1; b) node 2 ; c) node 3 and d) node *t*. (Assume for each case that the associated node is the highest priority node and, thus, ignore the possible consumption at other nodes.)

References

1. Endrenyi, J., *Reliability modeling in Electric Power Systems,* J.Wiley & Sons, New York (1978)
2. Barlow, R., Proschan, F., *Mathematical Theory of Reliability,* J.Wiley & Sons, New York (1965)
3. *Issues of the Mathematical Theory of Reliability*, (in Russian), Ed. Gnedenko, B.V., Radio i Svjaz, Moscow (1983)
4. Buzacott, J. A., Network approaches to finding the reliability of repairable systems, *IEEE Trans. Reliab.,* **R-19** (1970), pp. 140–146.
5. Singh, C., Billinton, R., A new method to determine the failure frequency of a complex system, *Microel. Reliab.,* **12** (1972), pp. 459–465.
6. Nahman, J., Failure frequency evaluation of complex systems using cut-set approach, *IEEE Trans. Reliab.,* **R-30**, (1981), pp. 352–356.
7. Samad, M.A., An efficient method for terminal and multi terminal pathset enumeration, *Microel. Reliab.,* **27** (1987), pp. 443–446.
8. Nahman, J., Minimal paths & cuts of networks exposed to common-cause failures, *IEEE Trans. Reliab.,* **R-41** (1992), pp. 76–80,84.
9. Singh, C., Billinton, R., *System Reliability Modelling and Evaluation* Hutchinson & Co., London (1977)
10. Allan, R. N., Billinton, R., De Oliveira, M. F., An efficient algorithm for deducing the minimal cuts and reliability indices of a general network configuration, *IEEE Trans. Reliab.,* **R-25** (1976), pp. 226–233.
11. Billinton, R., Allan, R., *Reliability Evaluation of Engineering Systems,* Plenum Press, New York (1992)

12. Nahman, J., Enumeration of minimal paths of modified networks' *Microel. Reliab.,* **34** (1994), pp. 475–484.
13. Nahman, J., Enumeration of minimal cuts of modified networks' *Microel. Reliab.,* **37** (1996), pp. 483–485.
14. Nahman, J., Exact enumeration of minimal paths of partitioned networks *Microel. Reliab.,* **34** (1994), pp. 1167–1176.
15. Philips, D. T., Garzia-Diaz, A., *Fundamentals of Network Analysis*, Prentice-Hall, Englewood Cliffs, N.J. (1981)
16. Christofides, N., *Graph Theory - An algorithmic Approach* , Academic Press, New York (1975).

5 Event Inspection Methods

The operation of many engineering systems depends in a very complex manner upon the failure modes of its components and cannot be simply modeled by a function graph. Sometimes it is even difficult to predict the impacts of various failures upon the dependability of the system before conducting serious and complex analyses of system postfailure states. Moreover, in the early design phase, a qualitative evaluation of the dependability of a system under consideration is often sought rather than a detailed one, as some of the subsystems are not completely conceived or defined. The next sections are devoted to some methods used in analyzing such systems.

5.1
FMECA and FTA Approaches

5.1.1
Failure Mode, Effects and Criticality Analysis (FMECA)

The basic principle of FMECA is to analyze the effects of each failure mode of every item of a system upon its operation [1,3,5]. The failure effects may be only qualitatively evaluated. The criticality analysis ranks the failures with regard to their *criticality number* attached to each system part failure mode. *The failure mode criticality number* is determined as

$$C_m = \beta_{pi} [1 - \exp(-\alpha_{pi}\lambda_p t)] \tag{5.1}$$

where β_{pi} is the conditional probability of the system losing its function given that mode i part p failure has occurred, α_{pi} is the probability that the failure of part p will be of mode i, λ_p denotes the total failure rate of part p and t is the operating or at-risk time of part p. At-risk time is the period of system operation during which the system is jeopardized by the failure of part p. The expression within the brackets in (5.1) yields the probability that mode i part p failure will occur during time period t. The *part criticality number* equals the sum of the part failure mode criticality numbers.

The application of FMECA requires experience and a good understanding and knowledge of the system operation and hardware. Special worksheets for the qualitative and criticality analysis have been devised and computerized [1].

This method is widely used in the design phase of various control systems, mission systems and processes. Very composite systems should be decomposed into subsystems to facilitate the FMECA approach.

As can be seen, the FMECA method indicates possible failure effects and most critical failures for a system. However, this method does not directly generate the dependability indices of a system.

5.1.2
Fault Tree Analysis (FTA)

The fault tree is a graphical representation of a sequence of events – causes and consequences that lead to the interruption of a certain function whose dependability is to be investigated. The analyst deduces which events devastate the system function being analyzed and what is the critical logical interrelationship between these events leading to such a consequence.

The fault tree is built by following some rules using the adequate symbols for logic operations and events. The most commonly used symbols are presented in Figs. 5.1 and 5.2 [1,2,3]. The *And Gate* indicates that the output event occurs if all input events have occurred. The *Or Gate* indicates that the output event occurs if any single event, or any combination of the input events, has occurred. The *Exclusive Or Gate* generates the output event if only a single among the input events has happened. The *Inhibit Gate* indicates that the input event directly produces the output event if the conditional input is satisfied. The *m out of n Gate* produces the output event if *m* or more of the input events have occurred.

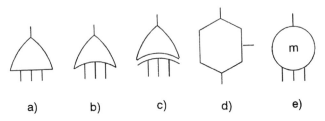

a)   b)   c)   d)   e)

Fig. 5.1 Logical symbols: *a) And Gate, b) Or gate, c) Exclusive Or gate, d) Inhibit Gate, e) m out of n Gate*

The event that cannot or deliberately should not be treated as a result of the occurrence of other events is *the base event*. The *output event* usually results from a combination of *input events*. An *undeveloped failure event* is an event not developed to its cause. A *significant undeveloped fault event* is one that requires further development for completion of the fault tree. A *condition event* opens the inhibit gate. The *transferred event* symbol indicates the transfer of the input event to another branch of the fault tree or the input of another branch of the fault tree into the branch under consideration.

The fault tree is built by beginning with the final failure event, *the top event*. The events causing the top event and their logical interrelationships are deduced. The

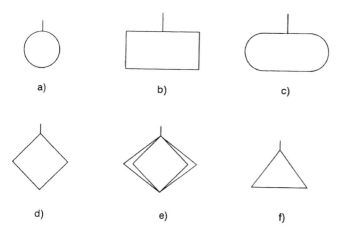

Fig. 5.2 Symbols for events: *a) base event, b) output event, c) condition event, d) undeveloped event, e) significant undeveloped event, f) transferred event*

composition of the tree gradually proceeds downward from output to input causal events until the base events are reached. The top event is the *root of the tree* while the input and output events correlated through the logical gates are *tree branches*. Base events are *leaves*. Logical gates can be considered as *nodes*.

Example 5.1
The construction of the fault tree will be demonstrated using the somewhat edited simple electric power system depicted in Fig. 5.3. Consumer *C* that is connected to busbar *S* is supplied from the power plant *E* via two overhead lines L_1 and L_2. Busbar *S* is also supplied by a local generating unit *G* that is connected to *S* through circuit breaker *B*. The fault tree should be composed to evaluate the dependability of the power supply to consumer *C*.

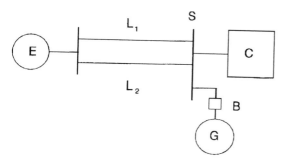

Fig. 5.3 Sample electric power system

The required fault tree is displayed in Fig. 5.4. The symbols used for the relevant failure events are:

C' - interruption of supply of consumer C

x - generating unit G and the supply system delivering the power from the remote power plant are simultaneously in failure states

z - generating unit G has short circuit failed and circuit breaker B has failed to open

y - both lines are simultaneously in a failure state or plant E is in a failure state

v - both lines are simultaneously in a failure state

E' - plant E is in a failure state

S' - busbar S is in a failure state

G' - generating unit G is in a failure state

B' - breaker B failed to open

L'_1, L'_2 - line L_1, L_2 is in a failure state

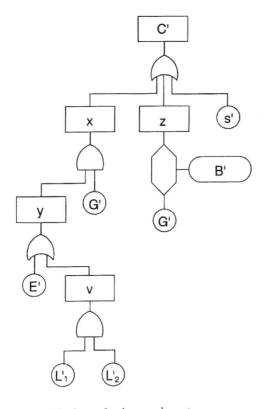

Fig. 5.4 Fault tree for the sample system

The top event – the interruption of supply of consumer C happens if x or z or S' occurs. Event z happens if a failure occurs in G *and if* circuit breaker B fails to open. In such a case the circuit breakers associated with the transmission lines, which are not presented for simplicity, will open to reduce the damage to the system and the supply of C will be interrupted. Event x occurs if both y *and G'* happen. Event y is caused by event E' *or* v, while v is caused if *both* lines are in failure states.   □

The expressions for the unreliability and unavailability of the system in conducting the function modeled by the fault tree may be derived by applying the general expressions for the probability of the union and of the overlapping of events. The probability that at least a single event among events e_i, $i = 1,..., m$, will occur is determinable using (4.37). This relationship yields the probability of the output event of an *Or Gate* with e_i being the probabilities of the input events. These events may be the unreliabilities or unavailabilities depending on what information is required. Both the time-specific and steady-state unavailabilities can be analyzed.

The general expression for the probability of overlapping of events e_i, $i=1,..., m$, is, as is known from the theory of probability [6],

$$\Pr\left\{ \prod_{i=1}^{m} e_i \right\} = \Pr\{e_1\}\Pr\{e_2|e_1\}\Pr\{e_3|e_1e_2\}\cdots\Pr\left\{ e_m \middle| \prod_{i=1}^{m-1} e_i \right\} \tag{5.2}$$

Eq. (5.2) gives the probability of the output event of an *And Gate* based upon the probabilities of input events.

Expressions (4.37) and (5.2) are valid for mutually dependent input events too. For independent input events (5.2) converts to a simple product of the corresponding probabilities.

Example 5.2
By applying (4.37) and (5.2) to the previous example we obtain

$$\Pr\{C\} = \Pr\{x\} + \Pr\{z\} + \Pr\{S'\} - \Pr\{x|z\}\cdot\Pr\{z\} - \\ - \Pr\{z\}\cdot\Pr\{S'\} + \Pr\{S'\}\cdot\Pr\{x|z\}\cdot\Pr\{z\} \tag{5.3}$$

Events x and z are mutually dependent as both of them are associated with event G'.
By inspecting Fig. 5.4 we deduce that

$$\Pr\{x\} = \Pr\{y\}\cdot\Pr\{G'\} \tag{5.4}$$

$$\Pr\{z\} = \Pr\{B'|G'\}\cdot\Pr\{G'\} \tag{5.5}$$

From (5.5) it can be seen that if z has happened it implies that G' has happened. Consequently, bearing in mind (5.4),

$$\Pr\{x|z\} = \Pr\{y\} \tag{5.6}$$

For events y and v we have

$$\Pr\{y\} = \Pr\{E'\} + \Pr\{v\} - \Pr\{E'\}\cdot\Pr\{v\} \tag{5.7}$$

$$\Pr\{v\} = \Pr\{L_1'\}\cdot\Pr\{L_2'\} \tag{5.8}$$

As the unreliabilities or unavailabilities of system components are presumed to be known base inputs, as well as probability $\Pr\{B\,|\,G'\}$, by applying the above expressions in reverse, bottom-up sequence we calculate the required probability of top event C', i.e. the unreliability or unavailability of the sample system in performing the function analyzed. □

The approach outlined is effective when applied to relatively simple systems or to systems having very reliable components so that higher order coincidences of failure events are extremely rare events and may be ignored. More complex systems can be analyzed by applying the *minimal cut* concept. The cut for a fault tree is a set of failure events whose overlapping causes the interruption of the system function under consideration. A minimal cut is such a set of events of which no subset is a cut. The procedure in deducing the minimal cuts from a fault tree is rather simple and easily programmable. If the output event of an *Or Gate* is a cut then all input events are cuts. If the output event of an *And Gate* is a cut then the input events constitute the set of events defining this cut. These two rules are successively applied beginning from the top event and gradually downward until the base events are reached. The cuts built from base events we call *terminal cuts* as they cannot be further developed. The terminal cuts determined along this procedure are afterwards checked for being minimal.

Example 5.3
Let us consider again the system displayed in Figs.5.3 and 5.4. In the first step we deduce that events x, z and $S'$are cuts. As $S'$is a base event it is a first order minimal cut. Event z is a second order cut whose elements are events $G'$and B'. As these two events are base events, this cut is a terminal cut and the analysis of the corresponding branch of the fault tree is ended. Event x is a cut whose elements are events $G'$and y. As y is caused by E'or v, we deduce from the previous cut two cuts, one with elements $G'$and $E'$and the other with elements $G'$and v. As $G'$and $E'$are base events they form a terminal cut. Event v is caused by an *And Gate* from events L_1' and L_2'. Consequently cut $G'v$ develops into $G'L_1'L_2'$. As the elements of this cut are base events it is a terminal cut. By inspection we conclude that all terminal cuts are minimal cuts.

The dependability indices are obtainable from minimal cuts by following the procedure already described in Section 4.2 for network minimal cuts. □

5.2
Failure Event Set Enumeration

Some engineering systems depend on the states of their components in such a complex manner that the impacts of various component states upon the functions of the system cannot be simply evaluated by inspection. For many systems a serious analysis should be conducted of post-failure system behavior bearing in mind the possible control and protection actions available for avoiding or reducing the undesired impacts. This is particularly the case for systems containing components with multiple failure states associated with various grades of functional deterioration. A typical example is an electrical power system which is composed of large sets of generating units and consumers interconnected through the transmission network with finite transfer capabilities. For such systems methods based upon enumeration of system states are applied. Some of the most effective of those approaches are now discussed.

5.2.1
Minimal Cuts

A general definition of *minimal cuts* may be introduced, being generally valid for dependability coherent systems [4]: a minimal cut is a set of degraded states of system components causing the deterioration of the system function under consideration; this set is characterized by the property that the transition of any set component state to any of its upgraded states restores the normal system function.

The probability of the overlapping of the events associated with a minimal cut yields alone the probability of system failure. We shall simply prove that. Let event E_i be the overlapping of component degraded states defined by minimal cut i. We denote by H the event that the remaining system components are in a specified combination of states. The probability that the system function has deteriorated if both aforementioned events have happened equals

$$\Pr\{E_iH\} = \Pr\{H|E_i\} \cdot \Pr\{E_i\} \tag{5.9}$$

Bearing in mind that event E_i causes a deterioration of system function irrespective of the remaining system component states, the probability

$$\Pr\{E_iH_c\} = \Pr\{H_c|E_i\}\Pr\{E_i\} \tag{5.10}$$

is also associated with a degraded system state. H_c denotes the event complement to H. By summing (5.9) and (5.10) we obtain

$$\Pr\{E_iH\}+\Pr\{E_iH_c\}=\left(\Pr\{H|E_i\}+\Pr\{H_c|E_i\}\right)\Pr\{E_i\}=\Pr\{E_i\} \tag{5.11}$$

As can be seen from (5.11), $\Pr\{E_i\}$ yields the probability of system state deterioration for any H. Clearly, the total probability of system deterioration is determined by applying (4.37) to all system minimal cuts.

Minimal cuts are determinable by successive enumeration of degraded states of system components. In the first step all degraded states of all components are checked for being first order minimal cuts. The second order minimal cuts are determined by inspecting the impacts of all combinations of degraded states of pairs of system components. To determine the third order minimal cuts, the effects of combinations of degraded states of triples of system components are inspected, etc. The enumeration is terminated after all minimal cuts up to a specified order are deduced, which is selected according to the accuracy required of the dependability analysis. In the enumeration procedure all cuts containing other cuts as subsets are discarded as they are not minimal by definition. This substantially reduces the number of combinations which have to be considered.

Example 5.4
Consider the sample system in Fig. 5.3. The enumeration of failure events to determine the minimal cuts of the system is conducted in the following sequence:

$$E'_1, L'_1, L'_2, (S'), G'_1, B'_1, E'L_1, E'L'_2, (E'G'), E'B'_1, L'_1L'_2, L'_1G'_1, L'_1B'_1, L'_2G'_1,$$

$$L'_2B'_1, (G'B'), E'L'_1L'_2, E'L'_1B'_1, E'L'_2B'_1, (L'_1L'_2G'), L'_1L'_2B'$$

The events that are minimal cuts are indicated within parentheses. They are the same as those deduced from the fault tree in the preceding example. B' denotes the stuck to open event. The only unfavorable effects of this event are associated with G'_1, as explained previously. It is presumed that all components have only one degraded state – the total failure state. As observed, the events that are minimal cuts are not combined with other events as such combinations would generate nonminimal cuts. The total number of combinations of failure events for the sample system equals $2^6 - 1 = 63$. However, only 21 combinations had to be analyzed to deduce all minimal cuts.

□

The method based upon minimal cuts makes it possible to determine all time-specific or steady-state system dependability indices by following the same approach as shown for networks in Chapter 4. Nevertheless, if we are interested not only in whether the system is normally operating but also in what is the grade of its deteriorations, the system state-space enumeration method should be used.

5.2.2
System State-Space Enumeration

The enumeration method inspects system states for dependability by following a sequential procedure which is basically from more to less probable states. The total sum of the probabilities of system states is 1 if all possible states are enumerated.

The procedure includes the following steps:

1. A system state, say k, is selected and its probability is determined as

$$P_k = \Pr\{ \prod_{i=1}^{n} e_{ki} \} \tag{5.12}$$

with e_{ki} denoting the event of component i being in a specified state and n designating the total number of system components. For independent components (5.12) converts into a simple product of component states probabilities.

2. The operation of the system in the selected state is analyzed using corresponding tools designed for the type of system under consideration in order to determine whether the analyzed system function(s) has (have) deteriorated and if so, to what degree. If the answer is negative the procedure continues with step #4. Otherwise, the probability of system being in a failure state is calculated as

$$P_f = P_{fs} + P_k \tag{5.13}$$

Index s labels the corresponding result obtained while conducting steps #1 and #2 for states 1 to $k-1$.

3. The expected value of the relevant system performance index is determined as

$$E\{V\} = E\{V\}_s + P_k V_k \tag{5.14}$$

where V_k is performance index value in state k. Continue with step #5.

4. The probability of sound system operation is determined as

$$P_g = P_{gs} + P_k \tag{5.15}$$

where P_{gs} is the sound system operation probability calculated up to state $k-1$ by following this procedure.

5. If $k < N$, with N being the total number of system states, the procedure continues with step #1 for state $k+1$. Otherwise, the calculation is terminated.

In step #3 we can simultaneously calculate expected values for several different performance indices.

The enumeration procedure is commonly applied for the steady-state dependability analysis. However, it can be used for time-specific analysis too, if required. In this case, it is presumed that the time-specific probabilities of component states are known. In the steady-state analysis, the enumeration procedure is usually conducted up to a preadapted level of coincidence of component degraded states. The main reason for that is the extremely large number of states a composite system might have. Fortunately, there is a simple way to determine the errors in evaluation of system availability and unavailability introduced by the state-space truncation [4].

Let U^* and A^* be the steady-state unavailability and availability of the system (probabilities P_f and P_s in the previous enumeration procedure) calculated for the truncated state space. As the exact values of unavailability and unavailability add to 1, the total probability of discarded states equals

$$\epsilon = 1 - (A^* + U^*) \tag{5.16}$$

It is obvious that the exact availability A and unavailability U are within the following intervals

$$A^* \leq A \leq A^* + \epsilon \tag{5.17}$$

$$U^* \leq U \leq U^* + \epsilon \tag{5.18}$$

As the enumeration goes from better to worse states it is reasonable to presume that the most of the discarded states are system failure states. Therefore,

$$U \approx U^* + \epsilon \tag{5.19}$$

is a good approximation. Clearly, the results calculated can be checked for accuracy at each step of the calculation procedure using (5.16), which provides a guidance to when to stop.

Example 5.5
The states of the system in Fig. 5.3 should be enumerated in the following sequence:

EL_1L_2SBG, $E'L_1L_2SBG$, EL'_1L_2SBG, $EL_1L'_2SBG$, $EL_1L_2S'BG$, $EL_1L_2SB'G$,

EL_1L_2SBG', $E'L'_1L_2SBG$, $E'L_1L'_2SBG$, $E'L_1L_2S'BG$, $E'L_1L_2SB'G$,...,etc.

It is worth noting that system states are defined by states, good or degraded, of all components. The total number of these states equals $2^6 = 64$. In searching for system minimal cuts, only degraded component states are taken into consideration.

□

5.2.3
Kronecker Algebra Application

In some cases the application of Kronecker algebra offers an easily programmable procedure for enumerating system states [4]. We shall consider the system consisting of several source units each of which may be in several capacity states characterized by various degrees of degradation. The problem to solve is to determine the system available capacity states and the probability of these states.

Unit k is defined by the diagonal matrix of possible unit capacity states

$$[x_k] = \text{diag}\{x_{1_k} \ ... \ x_{n_k}\} \tag{5.20}$$

and by the column vector of the associated probabilities

$$p_k = [p_{1_k} \ ... \ p_{n_k}]^T \tag{5.21}$$

It is presumed that unit k has n_k distinct available capacity states. The system states capacities are obtainable as

$$[x] = \bigoplus_{k=1}^{n} \text{diag}\{x_{1_k} \ ... \ x_{n_k}\} \tag{5.22}$$

The associated probabilities are calculated using the expression

$$[p] = \bigotimes_{k=1}^{n} [p_k] \tag{5.23}$$

Eq. (5.23) yields the probabilities of all possible system states (compare with Section 3.2) and the diagonal elements of the matrix in (5.22) are the associated system available capacities.

To demonstrate the method more clearly let us consider a system consisting of two source units. Unit 1 has three capacity states and unit 2 has two capacity states. By applying (5.21) and (5.22) we obtain

$$[p] = \begin{bmatrix} p_{1_1} \\ p_{2_1} \\ p_{3_1} \end{bmatrix} \otimes \begin{bmatrix} p_{1_2} \\ p_{2_2} \end{bmatrix} = \tag{5.24}$$

$$= \begin{bmatrix} p_{1_1}p_{1_2} & p_{1_1}p_{2_2} & p_{2_1}p_{1_2} & p_{2_1}p_{2_2} & p_{3_1}p_{1_2} & p_{3_1}p_{2_2} \end{bmatrix}^T$$

$$[x] =$$

$$= \text{diag}\{x_{1_1}+x_{1_2} \ \ x_{1_1}+x_{2_2} \ \ x_{2_1}+x_{1_2} \ \ x_{2_1}+x_{2_2} \ \ x_{3_1}+x_{1_2} \ \ x_{3_1}+x_{2_2}\} \tag{5.25}$$

As may be seen from (5.24), the probabilities in [p] are not listed in descending order of magnitude. If necessary, this can be done by applying well known sorting routines. However, the elements in (5.25) should be then accordingly rearranged.

Example 5.6
Let the source units have the following parameters

$$[p_1] = [0.85 \ 0.10 \ 0.05]^T, \quad [x_1] = \text{diag} \{30 \ 15 \ 0\},$$

$$[p_2] = [0.95 \ 0.05]^T, \quad [x_2] = \text{diag} \{15 \ 0\}.$$

Elements of $[x_1]$ and $[x_2]$ are unit available capacities in possible states expressed in corresponding units of measure.
By applying (5.22) and (5.23) we calculate for the system

$$[p] = [0.8075 \ 0.0425 \ 0.095 \ 0.005 \ 0.0475 \ 0.0025]^T,$$

$$[x] = \text{diag} \{45 \ 30 \ 30 \ 15 \ 15 \ 0\}.$$

As may be observed, matrix [x] has two pairs of identical elements. The states associated with the same capacity level may be merged into a single state by summing their probabilities. If done so, reduced forms of system parameters are obtained

$$[p_r] = [0.8075 \ 0.1375 \ 0.0525 \ 0.0025]^T,$$

$$[x_r] = \text{diag} \{45 \ 30 \ 15 \ 0\}.$$

Such reduced forms may be further combined with remaining system components if any. □

Example 5.7
Let us assume that unit 2 in *Example 5.6* is a consumer having the same parameters as before. Elements of $[x_2]$ are amounts of production demanded of source unit 1. The unavailability of the system in supplying the consumer demands are to be calculated. No reserve capacity margins are foreseen.
The possible system states are the same as in *Example 5.6* and calculated in the same way.
The difference between the available and demanded capacities is

$$[y] = [x_1] \oplus (-[x_2])$$

Using the data for $[x_1]$ and $[x_2]$ we obtain

$$[y] = \text{diag} \{15 \ 30 \ 0 \ 15 \ -15 \ 0\}$$

Obviously, the fifth state is a deficiency state which is expected with probability 0.0475 as it follows from the system state probability vector. This probability is the steady-state unavailability of the system. Should any capacity margin be required from the system, the third and the sixth system states would be found unsatisfactory too and their probabilities should be added to the probability of the fifth state to obtain the system unavailability.

Let us assume that the sample considered is an elementary electric power system and that capacities and demands are expressed in MW. As any steady-state probability may be understood as the relative cumulative duration of the associated state during a review period, say a year, then the expected energy not supplied to the consumer annually (EENS) equals, in our example

$$EENS = 0.0475 \times 8760 \times 15 = 6241 \text{ MWh}$$

□

Problems

1. Consider an electrical power circuit breaker having the following major failure modes: $a)$ nonoperable mechanical gear, $b)$ slight overheating of main contacts under rated load conditions, $c)$ incapability of interrupting some specific fault currents (e.g. short line faults). Let: $\beta_{pa} = 0.2$, $\beta_{pb} = 0.2$, $\beta_{pc} = 1.0$, $\alpha_{pa} = 0.8$, $\alpha_{pb} = 0.15$, $\alpha_{pc} = 0.05$ and $\lambda = 0.04$ fl./yr. Determine the criticality numbers for all failure modes and the breaker criticality number for one year.

2. Consider a system composed of three identical units. One unit is operating while the remaining two units are in cold standby and equipped with imperfect (unreliable) switches. Construct the Fault Tree diagram for this system.

3. Determine the minimal cuts for the system in *Problem #2 a)* from the associated Fault Tree, *b)* by enumeration.

4. Consider the sample system in Fig. 5.3. Let the capacity of the remote power plant be 500 MW and that of the local generating unit 200 MW. The load demand of the consumer is 500 MW throughout the whole year. The load transmission capacity of a line equals 300 MW. By applying the system state-space enumeration method calculate the steady-state unavailability of the system in covering the full consumer demand and the expected energy not supplied to the consumer annually. The unavailabilities of system components are: $U_E = 0.95$, $U_G = 0.85$, $U_{L1} = U_{L2} = 10^{-3}$ $U_C = 10^{-5}$, $U_B = 0.01$.

5. Presume that all elements of the system in Fig. 5.3 are reliable except the generating plants. Using the Kronecker algebra determine the steady-state unavailability of the system in covering the full consumer demand and the expected energy not supplied to the consumer annually. Compare the results obtained with those calculated in *Problem #4*.

References

1. O'Connor, P. D. T., *Practical Reliability Engineering*, J.Wiley & Sons, New York (1991)
2. Fussel, J. B., Fault Tree Analysis – Concepts and Techniques, in: *Generic Techniques in Systems Reliability Assessment,* Ed. Henley, E.J., Lynn, J.W., Noordhoff, Leyden (1976), pp. 133–162.
3. Billinton, R., Allan R. N., *Reliability Evaluation of Engineering Systems* Plenum Press, New York (1992)
4. Nahman, J., *Methods for Dependability Analysis of Electric Power Systems* (in Serbian,) Naučna Knjiga, Belgrade (1992)
5. Villemeur, A., *Reliability, Availability, Maintainability and Safety Assessment,* Vols. 1 & 2, J.Wiley & Sons, New York (1992)
6. Papoulis, A., *Probability, Random Variables, and Stochastic Processes,* McGraw-Hill, New York (1984)

6 Finite-Term Dependability Prediction

In practical applications, it is sometimes required to predict the behavior of a system in the near future [1-4]. A typical example is the planning of the operating reserve in electric power systems where generating units are committed to maintain below tolerable limits the risk of insufficient generating capacity in meeting the expected load increase. The prediction is needed as there is a time delay associated with starting and loading the newly committed units. The dependability model should provide the basis for calculating the probability of various capacity states of generating units during a prospective time interval of, say, several hours in order to evaluate the risk of capacity deficiency situations during this planning period. Such an analysis enables the system operator to decide which new units and when have to be committed for security reasons. A finite-term analysis is particularly important for the so called *mission systems* which are designed to maintain their function during a relatively short time period – *mission time.*

It is clear that the time-specific dependability indices are easily determinable for systems which may be described by Markov models, as discussed in Chapter 3. However, there is a very wide set of engineering systems whose state residence times are not all exponentially distributed. The application of Markov models to such systems may lead to quite erroneous results for time-specific indices while the introduction of sets of fictitious states can substantially increase the number of system states and considerably enlarge the computational burden. We shall present a model for finite term dependability analysis, being an extension of the open loop sequential approach already described in Section 2.4 for two-state systems. The approach will be extended to apply to multi-state systems.

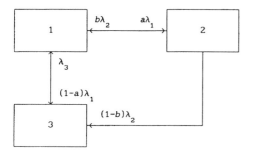

Fig. 6.1 Sample system with generally distributed state residence times

Consider for example a system having two failure modes, displayed in Fig. 6.1. It is presumed that a certain failure mode leads only to a degradation of system function while the other failure mode causes the total interruption of system service. $\lambda_k, k = 1,2,3$, are rates of transition from state k. As the residence times are generally distributed, these transition rates are dependent upon the times spent in the corresponding system states. Parameters a and b are probabilities of transitions from state 1 to state 2 and from state 2 to state 1, respectively, given the transitions from states 1 and 2 have occurred.

By following the general idea of the approach outlined in Section 2.4, the system under consideration may be modeled by a sequence of repetitive cycles, each cycle associated with a new restoration of the sound system state as well as of the partial and complete failure states (Fig. 6.2).

The probability of the system being in state k equals the sum of the probabilities of this state in the series of cycles. It is clear that the exact answer would be obtained with an infinite number of system cycles. Fortunately, for finite time analysis, only a few cycles should suffice for a fair assessment of system behavior. Furthermore, the error bounds may be relatively simply determined, as shown hereafter. The point worth noting is that the probabilities of system states in a cycle depend upon the probabilities of the preceding states only and may be recurrently determined for all cycles as we shall show for the sample being studied.

Define the *survival function*

$$S_k(z) = \exp\left(-\int_0^z \lambda_k(u)du\right) \qquad k=1,2,3 \tag{6.1}$$

which is the probability that the system will not abandon state k if it has resided in this state for time period z.

With regard to Fig. 6.2 we have

$$p_{1.1}(t) = S_1(t) \tag{6.2}$$

$$p_{2.1}(t) = \int_0^t p_{1.1}(x)a\lambda_1(x)S_2(t-x)dx \tag{6.3}$$

$$p_{3.1}(t) = \int_0^t p_{1.1}(x)(1-a)\lambda_1(x)S_3(t-x)dx +$$

$$+ \int_0^t p_{2.1}(x)(1-b)\lambda_2(x)S_3(t-x)dx \tag{6.4}$$

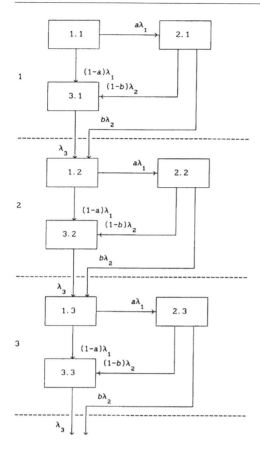

Fig. 6.2 Open loop cyclic state-transition diagram

It may be seen that, by definition of the survival function, $S_1(t)$ is the reliability of the system (see Section 1.1). The three first terms of the integrand in (6.3) multiplied by dx yield the probability that state 2 has been encountered at time instant x while the survival function term gives the probability that this state will not be abandoned before t. The meaning of the terms in (6.4) are similar and self-explanatory.

The probability of the sound state in the second cycle equals

$$p_{1.2}(t) = \int_0^t [p_{2.1}(x)b\lambda_2(x) + p_{3.1}(x)\lambda_3(x))] \, S_1(t-x)dx \tag{6.5}$$

By following the same reasoning as above we may write for the probabilities of states in cycle K

$$p_{1.K}(t) = \int_0^t [p_{2.K-1}(x)b\lambda_2(x) + p_{3.K-1}(x)\lambda_3(x)] \, S_1(t-x)dx \tag{6.6}$$

$$p_{2.K}(t) = \int_0^t p_{1.K}(x) a \lambda_1(x) S_2(t-x) dx \qquad (6.7)$$

$$p_{3.K}(t) =$$

$$= \int_0^t [p_{1.K}(x)(1-a)\lambda_1(x) + p_{2.K}(1-b)\lambda_2(x)] \, S_3(t-x) dx \qquad (6.8)$$

It is obvious that

$$p_i(t) > \sum_{K=1}^N p_{i.K}(t) , \qquad i=1,2,3 \qquad (6.9)$$

which means that the expression on the r.h.s. of (6.9) is the lower bound for $p_i(t)$ if N is the last cycle taken into account.

The upper bound may be determined by following a simple procedure. Consider cycle N. Do the following:

a) Convert the state whose probability upper bound should be determined, say state i, into an absorbing state by omission of $S_i(t-x)$ in the corresponding probability expression.

b) All transitions outgoing from cycle N to cycle $N+1$ direct to state i .

c) Determine the probability of state i for cycle N modified according to steps a) and b) Let us denote the probability obtained as $p_{ai.N}(t)$.

It is clear that

$$p_{ai.N}(t) > \sum_{K=N+1}^\infty p_{i.K}(t) \qquad (6.10)$$

as the system has been modified to converge to state i.

Consequently, the upper bound for $p_i(t)$ is

$$p_i(t) < \sum_{K=1}^{N-1} p_{i.K}(t) + p_{ai.N}(t) \qquad (6.11)$$

For illustration, Figs. 6.3 to 6.5 display the terminal cycle modified for determining probabilities $p_{aNi}(t)$ for the sample system states.

In addition, two points are worth noting. System state probabilities in cycle K are the probabilities that the corresponding states are encountered K times which is helpful information in many cases. Furthermore, the approach discussed makes it

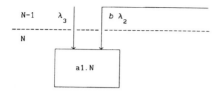

Fig. 6.3 N cycle transition diagram for calculating $p_{a1.N}(t)$

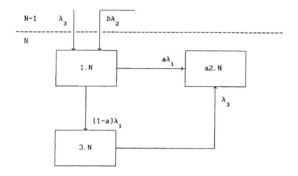

Fig. 6.4 N cycle transition diagram for calculating $P_{a2.N}(N)$

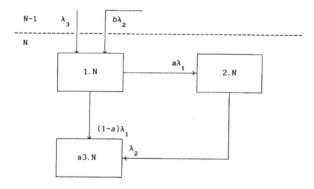

Fig. 6.5 N cycle transition diagram for calculating $p_{a3.N}(t)$

possible to account for the variation of state residence time distributions with the number of transitions. This might substantially enhance the model for some practical cases.

In the recurrent expressions for state probabilities some of the associated integrals cannot be solved in analytical form. However, all state probabilities may be calculated for various t by applying numerical integration.

The general form of the probability expression is

$$p(t) = \int_0^t F_1(x) F_2(t-x) dx \qquad (6.12)$$

Let us split the integration interval $(0,t)$ into q equal time steps Δt yielding

$$t = q\Delta t \tag{6.13}$$

Then, by applying the trapezoidal integration formula [5], we obtain

$$p(q\Delta t) \approx \frac{\Delta t}{2} [F_1(0)F_2(q\Delta t) + F_1(q\Delta t)F_2(0)] +$$
$$+ \Delta t \sum_{j=1}^{q-1} F_1(j\Delta t)F_2((q-j)\Delta t) \tag{6.14}$$

The second term in (6.14) is zero for $q<2$. Expression (6.14) is valid for all $q>0$, i.e. t. From (6.12) it is clear that $p(0)=0$. It is worth noting that $p(q\Delta t)$ is the part of the integrand in the formula for the probability of the state to which the state under consideration may transit (succeeding state), that is of the same form as (6.12). Eq. (6.14) provides all data needed on the preceding state probability to be inserted in (6.14) when applied to the succeeding state. It is implied that the same Δt is adopted throughout the entire calculation procedure.

Example 6.1
Consider the system displayed in Fig. 6.1. Let the sound operation duration be exponentially distributed with $\lambda_1=1$ fl./10^4 h and $a=0.8$. The partially degraded state residence time is exponentially distributed with $\lambda_2=1$ trans./10 h and $b= 0.3$.

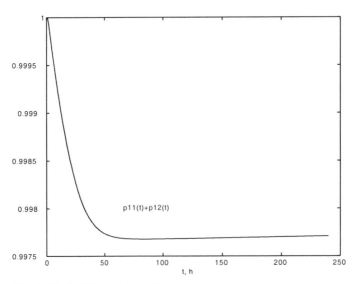

Fig. 6.6 Probability $p_1(t) \approx p_{1.1}(t) + p_{1.2}(t)$

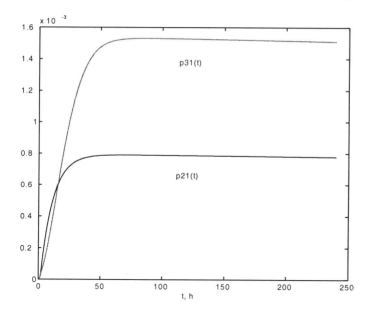

Fig. 6.7 Probabilities $p_{2.1}(t)$ and $p_{3.1}(t)$

The total failure repair duration is estimated to be Weibull distributed with $v = 0$, $\alpha=23$ h and $\beta=2$. Fig. 6.6 displays the probability of the system being in the sound state at any instant during a 240 h time period, approximately determined by using the results of the first two calculation cycles. It is presumed that the system is sound at $t = 0$. For integration, $\Delta t=1$ h has been adopted. Fig. 6.7 depicts calculated probabilities $p_{2.1}(t)$ and $p_{3.1}(t)$.

Problems

1. Assume that the total failure state residence time of the system in *Example 6.1* is exponentially distributed with $\lambda_3=1$ trans./23 h. Using the Markov model calculate $p_1(t)$ and compare the results obtained with those in *Example 6.1* during a 24 h time period (refer to Section 3.1).

2. For the system in *Example 6.1* determine the probability of being in state 2 during the 240 h period, using the results of the first two cycles. It is presumed that the system is in state 1 at $t=0$.

3. Solve Problem #2 under assumption that the system is in state 3 at $t=0$.

4. Calculate the higher bounds of the state probabilities of the system in *Example 6.1* during the 240 h period if only the first cycle is considered.

References

1. Shooman, M.L., *Probabilistic Reliability: An Engineering Approach*, McGraw-Hill, New York (1968)
2. Endrenyi, J., *Reliability Modeling in Electric Power Systems*, J.Wiley & Sons, New York (1978)
3. Patton, A. D., Short-term reliability calculation, *IEEE Trans. Power Appar. Syst.*, **PAS-89** (1970), pp. 509–513
4. Patton, A. D., A probability method for bulk power system security assessment II – Development of probability models for normally operating components, *IEEE Trans. Power Appar. Syst.*, **PAS-91** (1972), pp. 2480–2485
5. Volkov, E. A., *Numerical Methods*, Mir Publishers, Moscow (1986)

7 Simulation

7.1
Fundamentals

7.1.1
General

System dependability analysis may be conducted by simulating its operation and the randomness in the behavior of system components. The general approach comprises two main phases:

a) The possible states of system components are generated at random in accordance with their probability, by accounting for all interrelationships among components and technical or/and procedural constraints.
b) The system state determined in the preceding step is analyzed with regard to all relevant technical merits to evaluate system dependability in this state.

By conducting the aforementioned calculation, a simulation trial of possible system behavior is obtained. If these steps are repeated many times or continuously for a long time period, data samples are obtained for all system indices of interest which may be processed by applying statistical methods.

The simulation approach has some advantages and drawbacks when compared with the analytical approach.

The advantages are:
- General distributions of system component state residence times can be modeled.
- Deterministic rules or/and procedures may be taken into account.
- Correlations among various events and interdependence of component states may be incorporated.
- The entire history of system operation up to the time period of interest can be accounted for, which is crucial for systems with limited resources.
- Probability distributions for all variables of interest may be obtained in many cases, rather than only mathematical expectations.

The following drawbacks of the simulation approach should be mentioned:
- The accuracy of the results is limited and maximum errors introduced are not always determinable.
- A large number of samples or extremely long simulation time periods are needed to obtain credible results which implies a very substantial use of computer time. With nowadays computers this is not usually a very serious problem.

– If the sensitivity of the dependability indices to certain parameters is to be assessed, the simulation flow should be repeated by using the same sequence of random numbers generating system component states. Otherwise, the comparison would be quite ambiguous. However, even if this approach is used, the comparison will remain uncertain to a degree unless the effects of the parameter under consideration upon system performance are considerable.

Bearing the above in mind, the analyst should decide between the analytical and simulation method by considering the type of the problem to be solved. The following sections of this chapter will, we hope, facilitate such a choice. Sometimes the analytical and simulation approaches can be successfully combined.

7.1.2
Processing the Outcomes

Let us assume that a variable or index Z should be evaluated after a sequence of independent simulation trials. The arithmetic mean value of Z is

$$z_M = \frac{1}{n} \sum_{k=1}^{n} z_k \tag{7.1}$$

where z_k is the corresponding outcome in simulation k and n is the total number of trials.

From the central limit theorem it follows that for sufficiently large n [1]

$$\Pr\left\{ |E\{Z\} - z_M| \le \beta \frac{s}{\sqrt{n}} \right\} \approx \gamma \tag{7.2}$$

where $E\{Z\}$ is the mathematical expectation of variable Z and s is its standard deviation

$$s = \sqrt{\frac{1}{n-1} \sum_{k=1}^{n} (z_k - z_M)^2} \tag{7.3}$$

Probability γ, called the *confidence level*, and parameter β are correlated by the relationship

$$\gamma = \sqrt{\frac{2}{\pi}} \int_{0}^{\beta} \exp(-\frac{t^2}{2}) dt \tag{7.4}$$

From (7.2) it follows that, with probability γ,

$$z_M - \beta \frac{s}{\sqrt{n}} \leq E\{Z\} \leq z_M + \beta \frac{s}{\sqrt{n}} \tag{7.5}$$

Expression (7.5) defines the *confidence interval*.

Table 7.1 gives parameter β values for various γ. It is usually taken that (7.5) is valid if $n > 30$.

Table 7.1 Parameter β values

γ	0.80	0.85	0.90	0.95	0.995
β	1.281	1.440	1.645	1.960	2.801

As may be seen from (7.5), the maximum absolute error introduced by approximating the mathematical expectation by the arithmetic mean of outcomes obtained in the simulation process is inversely proportional to the square root of the number of simulations, with probability γ. This means that the convergence to the exact solution when n is increased is relatively slow.

As stated, the expressions given above are strictly valid for independent simulations. However, they may be used as approximations for dependent simulations too.

Sometimes, more information about Z is required besides the mathematical expectation and standard deviation. A more complete information provides the *relative frequency distribution* of Z which can be extracted from the simulation results. The following steps should be conducted to construct this distribution. The interval between the highest and lowest z_k value should be divided into a number of subintervals of the same size. This number may be chosen by applying (1.25) or (1.26). The subintervals are chosen so that the midpoints of subintervals coincide with a z_k value. The subinterval boundaries should not coincide with any z_k value. The relative number of z_k falling into each subinterval has to be determined, based upon the total number of trials. It is obvious that the relative frequency distribution is an approximation of the pdf of Z.

Example 7.1

Assume that 50 independent trials have been conducted by simulation to determine the mean time between system failures (MTBF). The interval between the highest and lowest time between failures observed was 1200 h – 200 h = 1000 h. With regard to (1.26) we divide this interval in $50^{1/2} \approx 7$ subintervals whose length is $1000/7 = 142.9$ h ≈ 140 h. Table 7.2 yields the simulation results distributed among the chosen subintervals.

Table 7.2 Simulation results

Subintervals h	Number of trials n_k	Relative frequency $f=n_k/50$
200 – 340	3	0.06
340 – 480	8	0.16
480 – 620	11	0.22
620 – 760	10	0.20
760 – 900	7	0.14
900 – 1040	6	0.12
1040 – 1200	5	0.10

The relative frequency diagram is depicted in Fig. 7.1.

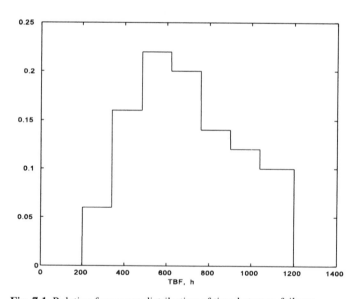

Fig. 7.1 Relative frequency distribution of time between failures

If the operation of the system depends on its behavior in the preceding time intervals, it should be simulated as a stochastic process respecting the sequence of events. The mean value of variable $Z(t)$ which is observed during the process is determinable as [2]

$$z_M = \frac{1}{T} \int\limits_0^T Z(t)dt \tag{7.6}$$

Providing simulation time interval T is long enough, z_M is taken to be equal to the mathematical expectation of Z. There are two possibilities to check if the selected T is appropriate. One may repeat the simulation process after changing the initial values of relevant quantities and compare the results obtained with the previous ones. If the difference is small, the result may be accepted as being correct. The other way is to prolong the simulation for a time interval ΔT and compare the results thus obtained with the previous ones.

It is implied that the process being considered is stationary and ergodic [2] and that, consequently, the expected value exists. More details on process modeling will be given in Section 7.4.

7.2
Generation of Uniformly Distributed Random Numbers

As will be shown in the following sections of this chapter, random variables associated with a system are modeled, both in steady-state and time-specific simulation, by successive generation of random numbers uniformly distributed within interval (0,1). There are several methods used by computers to produce a sequence of such numbers. Unfortunately, the numbers generated by computers are pseudo uniformly distributed as the sequences of successively generated numbers repeat cyclically. It is clear that the number generating software is the better the longer such a cycle is.

An approach commonly used by computers is the *linear congruental method* [3]. A sequence of nonnegative integers is produced by means of the recursive expression

$$x_{i+1} = (ax_i + c)(\text{mod } m) \tag{7.7}$$

The starting value x_0 is the *seed*, a is a constant *multiplier*, c is *increment* and m is *modulus*. The uniformly distributed number associated with x_i equals

$$R_i = \frac{x_i}{m} \tag{7.8}$$

As may be concluded from (7.7), x_{i+1} is the remainder of $ax_i + c$ divided by m. It means that for $ax_i + c < m$ we have

$$x_{i+1} = ax_i + c \tag{7.9}$$

For $c=0$ the *multiplicative congruental method* is obtained. It is interesting to mention that this method was used by the random number generator of IBM system 360, with $a=16807$ and $m=2^{31}-1$. For $m=2^b$, and $c=0$, the longest possible cycle equals 2^{b-2}. This cycle is achieved provided that x_0 is odd and $a=\pm 3 + 8k$ where k is an integer [3,4].

Example 7.2
Let us take $a=5$, $x_0=1$, $m=64$ and $c=0$. Expressions (7.7) and (7.8) yield the following sequence of random numbers (rounded to four significant digits)

$x_0=1$, $R_0=1/64=0.016$, $x_1=5\times1(\text{mod } 64)=5$, $R_1=5/64=0.078$,
$x_2=5\times5(\text{mod}64)=25$, $R_2=25/64=0.391$, $x_3=25\times5(\text{mod}64)=61$,
$R_3=61/64=0.953$, ...

7.3
Steady-State Dependability Indices

Consider the system composed of multistate mutually independent components discussed in Section 5.2. The steady-state dependability indices can be analytically determined using the enumeration or minimal cut approach. As shown in Section 5.2, the enumeration enables a simple error observation which is continuously descending in the course of the calculation flow. A drawback of this method lies in the complexity of selective enumeration programming, particularly if the minimal cut set approach is applied. The simulation approach implies much less programming effort as component states are generated by chance with no regard to any order. However, to achieve a satisfactory accuracy by simulation, a considerably higher number of system states should be generated than in the enumeration approach and the error introduced can be only statistically assessed, as shown in Section 7.1.

7.3.1
Generation of System States

Let us assume that a component may reside in s states. Let the associated probabilities be p_i, $i=1,...,s$. In simulation, the state of the component is selected by chance, by generating random number R uniformly distributed within interval $(0,1)$, with regard to the following expression

$$i = \begin{cases} 1 & \text{if} \quad 0 \le R \le p_1 \\ j & \text{if} \quad \sum_{k=1}^{j-1} p_k < R \le \sum_{k=1}^{j} p_k \end{cases} \qquad (7.10)$$

In (7.10) i is the index of the state selected by R.

Example 7.3
Consider a component having $s=3$ states with probabilities $p_1=0.8$, $p_2=0.15$ and $p_3=0.05$. With regard to (7.10), the component is in state 1 if $0 \le R \le 0.8$, in state 2 if $0.8 < R \le 0.95$ and in state 3 if $0.95 < R \le 1$.   □

To simulate a state of the system, random numbers should be generated for all system components in order to determine their states. Then, the system state generated in this simulation trial, say trial k, is analyzed with regard to all the properties of interest. Let us assume, for example, that the system availability is to be calculated. If the system is available, then the outcome of the simulation trial with regard to this index is $A_k=1$. Otherwise, $A_k=0$ should be stored.

The procedure described above should be repeated a large number of times until satisfactorily accurate results are obtained. This may be assessed by applying (7.5). An alternative approach is to define some stopping rules based upon the variations of mean values of relevant outcomes in a number of subsequent simulation trials. Providing this variation is small, it is taken that the calculation process has converged to the practically exact result.

Example 7.4
Consider a system composed of two series-connected components. Components may have two states: a sound state with probability 0.8 (steady-state component availability) and a failure state with probability 0.2 (steady-state component unavailability). The simulation approach should be applied to determine the steady-state system availability. The calculation flow is outlined in Table 7.3. The first column gives the total number n of simulation trials and the index number k of trials. The second and fourth columns are the uniformly distributed random numbers generated during the simulation process for consecutive trials. The third and fifth columns give the outcomes: observed component availabilities. According to (7.10), the availability of a component is 1 providing the generated random number lies in interval (0,0.8). Otherwise, the outcome is 0. The analyzed outcome for the system is system availability. As is known, this availability is $A=A_1A_2$. The simulation outcome in trial k is $A_k=A_{1k}A_{2k}$ with index k labeling the corresponding simulation trial. This result is listed in the sixth column. The seventh column lists the arithmetic mean values of system availability calculated from the data of column #6 by applying (7.1). The mean value is calculated after each simulation trial to demonstrate the simulation flow. The exact result is $A=0.8^2=0.64$. As may be seen, the result obtained after 15 trials is a rather crude assessment of the exact result. This is because the total number of trials is far too low.

In addition, it is worth noting that the result obtained after 14 trials is very close to the exact solution. Unfortunately, the analyst using simulation will be unaware of this fact generally, as his only means in estimating the accuracy of the results obtained is expression (7.5).

Table 7.3 Simulation flow for two-component system

n, k	R_1	A_{1k}	R_2	A_{2k}	A_k	A_M	C_k	C_M
1	0.125	1	0.000	1	1	1	150	150
2	0.679	1	0.999	0	0	0.5	50	100
3	0.328	1	0.998	0	0	0.333	50	83.3
4	0.832	0	0.242	1	0	0.250	100	87.5
5	0.002	1	0.071	1	1	0.400	150	100
6	0.368	1	0.530	1	1	0.500	150	108
7	0.246	1	0.281	1	1	0.571	150	114
8	0.752	1	0.390	1	1	0.625	150	119
9	0.764	1	0.049	1	1	0.667	150	122
10	0.359	1	0.064	1	1	0.700	150	125
11	0.000	1	0.111	1	1	0.727	150	127
12	0.386	1	0.973	0	0	0.667	50	121
13	0.689	1	0.703	1	1	0.692	150	123
14	0.498	1	0.878	0	0	0.643	50	118
15	0.233	1	0.139	1	1	0.667	150	120

Example 7.5   □

Suppose that the components in *Example 7.4* are parallel connected. Then, the system availability in simulation trial k equals $A_k=\max(A_{1k}, A_{2k})$. That is, the system is down only if both components are simultaneously down. By inspection of Table 7.3 we see that such a situation has not occurred in the first 15 simulation trials. This means that, based upon such a limited number of trials, it would be concluded that $A=1$ which is, of course, a false result. The exact solution is

$$A = A_1 + A_2 - A_1 A_2 = 0.96.$$ □

Example 7.6

Assume that the components in *Example 7.4* are parallel connected and that, transmission capacities of components 1 and 2 are, say, $C_1 = 50$ units and $C_2 = 100$ units. The expected capacity of the system should be determined using simulation.

The eighth column of Table 7.3 quotes system capacity observed in simulation trials while the ninth column lists the calculated arithmetic mean values. The exact expected system capacity is

$$C_k = A_1 A_2 (C_1 + C_2) + A_1 (1 - A_2) C_1 + A_2 (1 - A_1) C_2$$

$$= A_1 C_1 + A_2 C_2 = 120 \text{ units}$$

As may be observed from Table 7.3, after 15 simulation trials the exact solution has been obtained. Of course, the analyst will, again, be unaware of this fact.   □

7.4
Time-Specific Dependability Evaluation

7.4.1
Generation of Random State Residence Times

System components may transit from state to state as time flows. Since the state residence times are, generally, random varieties obeying a certain Cdf, in simulation, these times should be generated by chance for each simulation trial. As will be proved hereafter, random residence times may be obtained from the corresponding Cdf by means of random numbers R uniformly distributed within interval $(0,1)$ [4,5].

Consider the Cdf of a state residence time, depicted in Fig. 7.2. Let us generate R and determine the abscissa of the point on the Cdf corresponding to R. It is clear from Fig. 7.2 that $t \leq \tau$ if $R \leq F(\tau)$. Hence,

$$\Pr\{t \leq \tau\} = \Pr\{R \leq F(\tau)\} \tag{7.11}$$

If R is uniformly distributed within interval $(0,1)$, then

$$\Pr\{R \leq F(\tau)\} = F(\tau) \tag{7.12}$$

From (7.11) and (7.12) we obtain

$$\Pr\{t \leq \tau\} = F(\tau) \tag{7.13}$$

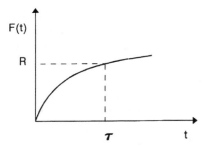

Fig. 7.2 Determination of state residence time

which means that if R is uniformly distributed within interval $(0,1)$ then generated τ values obey the Cdf of the state residence time.

According to Fig. 7.2, random τ is determinable by solving the equation

$$F(\tau) = R \tag{7.14}$$

A solution of (7.14) in analytical form can be obtained for the exponential residence time distribution. For

$$F(t) = 1 - \exp(-\lambda t) \tag{7.15}$$

the solution of (7.14) is

$$\tau = -\frac{1}{\lambda} \ln(1-R) \tag{7.16}$$

As $1 - R$ is also uniformly distributed within interval $(0,1)$, expression (7.16) may be replaced by a simpler one

$$\tau = -\frac{1}{\lambda} \ln(R) \tag{7.17}$$

In the case that the residence time is uniformly distributed within interval (a,b) (Section 1.3) the solution of (7.14) may also be obtained in analytical form

$$\tau = (b-a)R + a \tag{7.18}$$

However, for the majority of residence time Cdfs, Eq. (7.14) should be solved using numerical methods or available tabulated data.

7.4.2
System Reliability Evaluation

In many cases in practice the reliability of a system is of the principal interest as being the probability that the system will survive until a specified time. Characteristic practical examples of such type are mission time systems which have to be designed to successfully conduct a mission of precisely specified or estimated duration (satellite communications systems, aircraft systems, special military equipment and systems, etc.). The reliability of such systems may also be evaluated by simulation. By generating uniformly distributed random numbers the duration of system sound operation is determined in successive trials and compared with the specified mission time. If this time is exceeded in a trial, the system is assessed as reliable in this trial. The reliability of the system is obtained as the relative number of successful outcomes based upon the total number of trials.

Example 7.7

Consider a 1 out of 3: Good, nonrepairable system. The components have exponentially distributed lifetimes with $\lambda=5\times10^{-4}$ fl./h. The reliability of the system should be evaluated during 1000 h mission time.

With regard to (3.90) the exact analytical expression for system reliability is

$$R(t) = \sum_{k=0}^{1} \binom{3}{k} (1-\exp(-\lambda t))^k \exp(-(3-k)\lambda t)$$

For $t = 1000$ h we obtain $R(1000) = 0.657$, which is the exact result. Table 7.4 quotes the results obtained in ten simulation trials. In every trial the random number is generated for each component to determine, using (7.17), its lifetime in this trial. If at least two components are found to be up after 1000 h, the outcome is success.

Table 7.4 Results of the first ten trials

n, k	R_{1k}	τ_{1k}, h	R_{2k}	τ_{2k}, h	R_{3k}	τ_{3k}, h	relia-bility for k	relia-bility
1	0.598	1028	0.712	679	0.268	2633	1	1
2	0.241	2845	0.706	696	0.575	1107	1	1
3	0.848	330	0.110	4414	0.773	515	0	0.667
4	0.782	492	0.788	477	0.293	2455	0	0.500
5	0.238	2871	0.801	444	0.979	42	0	0.400
6	0.440	1642	0.133	4035	0.223	3001	1	0.500
7	0.634	911	0.077	5128	0.431	1683	1	0.571
8	0.190	3321	0.809	424	0.212	3102	1	0.625
9	0.798	451	0.307	2362	0.335	2187	1	0.667
10	0.319	2285	0.144	3876	0.696	725	1	0.700

This is indicated by setting 1 for the reliability of the system in the corresponding trial. Otherwise, the outcome is unsuccessful and zero should be inserted for system reliability. System reliability is calculated on the basis of generated trials using (7.1).

Fig. 7.3 shows the results obtained for 100 successive trials. As may be seen, after an oscillatory behavior at the beginning of the simulation the results are steadied, slowly converging to the exact result. The result obtained after 100 trials is 3.5% higher than the exact solution.

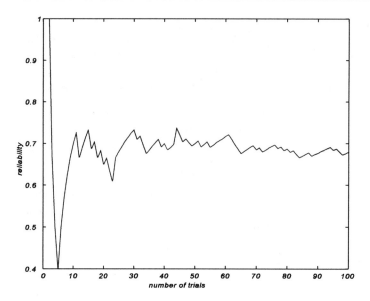

Fig.7.3 System reliability during first 1000 h obtained by simulation

7.4.3
Stochastic Processes

In many cases, the dependability of a system is affected by the sequence in which some events happened in the past. There are engineering systems with limited resources, such as water reservoirs and/or fuel stocks. These may have been consumed in some characteristic time intervals in the past which might affect the operational capability of the system in prospective periods [6]. Therefore, the operation of systems of the aforementioned type should be simulated as a stochastic process by strictly respecting the chronological flow of events.

Example 7.8
Consider a system consisting of a single component in service and of another component in cold standby. The following is presumed:
- The components have the same performances.
- The switch that sets the standby units in operation is perfect, i.e. it cannot fail and the time delay it introduces is short enough to be discarded.
- Components can fail during operation only.
- There is a single repair crew available.
- The components are repaired on the basis of the first failed first repaired rule.

The dependability analysis using the sequential simulation approach includes the following steps:

1. Generate R to determine the instant of failure of component 1, i.e. the time interval of sound operation, say m_1. After m_1 component 2 is set in service.
2. Generate R to determine repair duration of component 1, say r_1. Generate R to determine the period m_2 of sound operation of component 2.
3. Let us presume that $r_1<m_2$. In this case, the system is brought into the same state as in step #1, this time with component 2 in service and component 1 in cold standby. The further calculation flow proceeds principally as before, through steps #1 and #2 applied to the corresponding in-service and standby units.
4. If $r_1>m_2$, in step #3, the system is down for time period r_1-m_2 after which component 1 is set in service and the repair of component 2 is initialized.
5. Generate R to determine repair duration r_2 of component 2. Generate R to determine operation duration m_1 .
6. Let us presume that $r_2<m_1$. If so, the system is set in the same state as in step #1 and the calculation proceeds in the same way as before.
7. If $r_2>m_1$ in step #6, the system is down for time period r_2-m_1. After this period has expired the system encounters the same state as in step #2 and the calculation proceeds further by following the same sequence of steps as before.

$\square$

If the system unavailability is of interest, $Z(t)$ in (7.6) is 1 during down system states. In the remaining steps $Z(t)=0$ should be inserted.

If the down system states cause a certain cost, this cost should be measured in appropriate units $Z(t)$ during down states. The application of (7.6) yields, in this case, the cost per unit of time, say per year providing T is expressed in years [6]. Also, the relative frequency distribution of costs may be constructed based upon the set of cost data observed during the simulation flow. Such results cannot be obtained, for a general Cdf of sound operation and repair times, using an analytical approach.

Problems

1. A network is composed of two parallel branches series connected to a third branch. The steady-state availability of branches is 0.9. Estimate the steady-state availability of this network by simulation, using 30 trials. Compare this result with the result obtainable by the exact analytical approach. (Refer to *Examples 7.5* and *7.6* and Section 4.1)

2. Determine the confidence interval for the steady-state availability of the network in *Problem* 1 with probability 0.95.

3. If the transfer capacity of the parallel branches of the network in *Problem* 1 is 1.5 units each and of the third branch 2 units, calculate by simulation the expected (steady-state) transfer capacity of the network using 30 trials. Compare this result with the exact one obtained analytically. (*See Example 7.6*)

4. Assume that the sound service time of each of the parallel branches in *Problem 1* is uniformly distributed in interval (800 h,1500 h) while this time for the third branch is exponentially distributed with $\lambda = 1$ fl./2000 h. Determine by simulation the probability that the network will be operative after 1000 h given that the service started at $t=0$ with all branches being sound. (*See Example 7.7*)

5. Presume that the network in *Problem 1* is maintained by a single crew by applying the following rules: a) The repair of the third branch has to be undertaken immediately after its failure irrespective of the states of the parallel connected branches; b) The parallel connected branches are repaired by applying the first failed first repaired mode. The up times of branches are as in *Problem 4*. The repair duration for all branches is uniformly distributed in interval (5 h,10 h). (*See Example 7.8*)

References

1. Ross, S.M., *Introduction to Reliability Models*, Academic Press, New York (1997)
2. Bendat, J.B., Piersol, A.G., *Random Data – Analysis and Measurement Procedures,* J.Wiley & Sons, New York (1986)
3. Banks, J., Carson, J., S., *Discrete-Event System Simulation*, Prentice-Hall, Englewood Cliffs, New Jersey (1984)
4. Anders, G.J., *Probability Concepts in Electric Power Systems*, J. Wiley & Sons, New York (1990)
5. Billinton, R., Allan R.N., *Reliability Evaluation of Engineering Systems,* Plenum Press, New York (1992)
6. Nahman, J., Bulatović, S., Power system operating cost and committed generation capacity planning, *IEEE Trans. Power Systems*, **PWRS-11** (1996) pp. 1724–1729

8 Dependability Grading and Optimization

8.1
Optimization Concept

Inadequate system dependability performances may cause various undesirable consequences. Three main types of consequences might occur individually or combined:

a. Jeopardized human lives
b. Cost to the owner and /or user of the system
c. Inconvenience to the user

The improvement of the dependability indices implies extra cost for redundancy, preventive maintenance, spare parts, implementation of more expensive components, remote observation and/or control, automation, etc.

To optimize a renewable engineering system in long run, the cost-benefit model may be used. The optimum solution is that providing minimum total cost during the review period [1]

$$Z = \min_k \{C_k\} \tag{8.1}$$

where index k labels options under consideration.

C_k is the present worth of the total system cost during the review period, for option k

$$C_k = I_0 + \sum_{j=1}^{n} (C_{cj} + E_j)(1+p)^{-j} - (1+p)^{-n} I_n \tag{8.2}$$

The symbols in (8.2) denote:
I_0 - investment (capital) cost of the system
I_n - remaining system worth at the end of the review period
n - duration of the review period expressed in years
C_{cj} - service interruption cost for system users in year j
E_j - operating, preventive and corrective maintenance costs in year j
p - discount rate

The optimum solution has to satisfy all technical requirements which may be formally stated as

$$T \subseteq T_a \tag{8.3}$$

with T denoting the set of technical performance indices and T_a being the set of acceptable values of these indices.

Also, the dependability indices of the system should be acceptable which can be expressed as

$$D \subseteq D_a \tag{8.4}$$

where D is the set of relevant dependability indices and D_a is the set of acceptable values of these indices.

Cost C_{cj} may usually be expressed as

$$C_{cj} = \sum_i C_{ji} f_{ji} \tag{8.5}$$

with C_{ji} being the cost caused by service interruption i in year j and f_{ji} denoting the frequency of interruptions i in year j. Both C_{ji} and f_{ji} may change during the review period if the customers' structure and demands alter in time.

For mission time systems the total cost C_k could be taken to equal

$$C_k = I_0 + Q(T)C_M \tag{8.6}$$

where $Q(T)$ is the unreliability of the system defined as the probability that the system will fail before mission time T (Section 1.1). C_M is the cost of mission failure. If the mission failure cannot be assigned a monetary value, the optimal solution is the minimum capital cost option satisfying the corresponding performance criterion (8.4). This may be written in the form

$$Q(T) \leq \rho \tag{8.7}$$

with ρ being the prescribed acceptable failure risk.

Due to the great number and the complexity of relationships among various system parameters and variables it is usually not possible to apply an optimization technique to solve (8.1). Therefore, a limited number of optional solutions should be analyzed and compared to determine the best option. However, this implies that the best solution found depends on the set of options analyzed, which means that the designer selecting this set affects the result of optimization. Bearing the above in mind, it is recommended to ask several experienced professionals to propose their sets of feasible options and then to form the representative set as the union of these individual proposals.

Example 8.1
Consider an energy supplying unit whose up time is exponentially distributed with $\lambda = 1$ fl./1000 h. The number of hot standby supply units should be determined to limit to below $\rho=10^{-3}$ the risk of supply interruption before 500 h of service.

The aforementioned criterion is fulfiled if

$$(1-\exp(-\lambda t))^n \le \rho$$

with n designating the total number of supply units. From the latter expression we obtain

$$n \ge \frac{\log\rho}{\log(1-\exp(-\lambda t))}$$

For $t=500$ h we obtain $n \ge 3.21$ which yields $n = 4$. This means that the required number of standby units is 3.

□

The cost of the interruption of service depends upon the type of service and customers. The direct costs caused by the interruption of supply of various resources (water, electrical energy, gas, coal, etc.) can be assessed fairly well for industrial consumers and, with less accuracy, for commercial and residential consumers [2,3]. The same is the case with various plants and factories where postfailure scenarios may be precisely predicted and the losses implied evaluated. They generally depend on the severity of the impact caused and on the frequency and duration of the interruptions. However, *the loss of human life* may not be given a monetary value as life should be, for ethical reasons, treated as invaluable. Nevertheless, some researches have produced estimates of monetary equivalent of human life varying between US$ 0.2 million to US$ 4 million [4, p. 112]. The most appropriate way to estimate acceptable risk regarding humans is by referring to the involuntary and voluntary risks to which a person is exposed during his everyday activity. A list of these risks obtained by statistics is given below [5].

Voluntary risks	*Risk*
Smoking	300×10^{-5}
Drinking	7.5×10^{-5}
Playing football	4×10^{-5}
Car racing	120×10^{-5}
Rock climbing	14×10^{-5}
Car driving	17×10^{-5}
Motor cycling	2000×10^{-5}

Involuntary risks	*Risk*
Run over by vehicle	300×10^{-7}
Flood	22×10^{-7}
Earthquake	17×10^{-7}

Tornado	22×10^{-7}
Lightning	1×10^{-7}
Air crash	1×10^{-7}
Release from nuclear CS	1×10^{-7}
Influenza	2000×10^{-7}

A reasonable approach is to limit the risk of loss of life to a value below the lowest involuntary risk.

The inconveniences caused by service interruption are difficult to express in monetary equivalents. Consider, for example, the impacts caused to residential consumers by the interruption of the supply of electrical power. How should we quantify the impossibility of using elevators, air conditioning, radio and TV sets, personal computers, timers and other electrical appliances used for cleaning, cooking and house maintenance? How should we value the interruption of heating if no alternative means are available or the interruption of water supply provided by electrically driven pumps? How should we value the inadequate capacity of a telecommunication network introducing access time delays or phone call interruptions? The most reasonable way to evaluate such impacts is to define some criteria with regard to the relevant dependability indices of the engineering systems under consideration which should be enforced in both the design and the operating phase of this system. These criteria should be founded upon past experience, customer surveys and comparison with other systems of the same type. A promising approach in defining such criteria and dependability evaluation is to use fuzzy logic which makes it possible to incorporate linguistic evaluation of dependability performances [6]. This will be demonstrated in the subsequent sections.

8.2
Fuzzy Logic Based Dependability Evaluation

8.2.1
Fuzzy Sets

A fuzzy set $A(x)$ is defined as [7–9]

$$A(x) = \{x, \mu_A(x)\} \qquad x \in E \tag{8.8}$$

with $\mu_A(x)$ designating the membership grade of x in set $A(x)$ and E being the universal set. The highest possible membership grade is usually taken to be 1. $A(x)$ is a generalization of conventional crisp sets for which a member x either belongs to a set (its membership grade is 1) or does not belong to this set (its membership grade is 0). Fuzzy sets are capable of measuring linguistic terms such as large, low, acceptable, ... by attaching to each x which is to be evaluated, the membership grade in the set of large, low, acceptable ... values. Fig. 8.1 presents a triangular

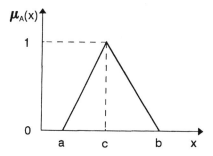

Fig. 8.1 Triangular membership function

membership function, for illustration. If x are real numbers, then $A(x)$ is a *fuzzy number* allowing us to represent the uncertainties in assessing a number, say an input parameter value.

The *support* of $A(x)$ (supp$\{A\}$) is the crisp set containing all $x \in E$ for which $\mu_A(x) > 0$. For $\mu_A(x)$ in Fig. 8.1 the supp$\{A\}$ are all $x \in (a,b)$.

The *kernel* of $A(x)$ (ker$\{A\}$) consists of elements x of $A(x)$ whose membership grade is 1. For Fig. 8.1 ker$\{A\} = c$.

The *height* of $A(x)$ (height$\{A\}$) is the highest membership grade. A fuzzy set is considered to be *normalized* if height$\{A\} = 1$ (which is the case in Fig. 8.1).

An α - *cut*, let us denote it as A_α, is a crisp set containing all $x \in E$ for which $\mu_A(x) \geq \alpha$. Obviously, $A_0 = \text{supp}\{A\}$.

Intersection $C(x)$ of $A(x)$ and $B(x)$ is constructed as (Fig. 8.2)

$$C(x) = A(x) \cap B(x)$$
$$\mu_C(x) = \min(\mu_A(x), \mu_B(x)) \quad x \in E$$

(8.9)

Union $C(x)$ of $A(x)$ and $B(x)$ is defined as (Fig. 8.3)

$$C(x) = A(x) \cup B(x)$$
$$\mu_C(x) = \max(\mu_A(x), \mu_B(x)) \quad x \in E$$

(8.10)

$\mu_c(x)$ is depicted in Fig. 8.3.

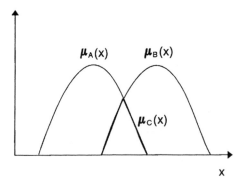

Fig. 8.2 Membership function of the intersection of two fuzzy sets

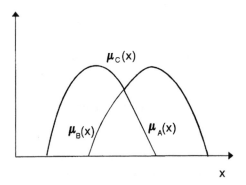

Fig. 8.3 Membership function of the union of two fuzzy sets

$A(x)$ is *equal to* $B(x)$ if and only if $\mu_A(x) = \mu_B(x)$ $x \in E$.

$B(x)$ is the *complement* to $A(x)$ if and only if $\mu_B(x) = 1 - \mu_A(x)$ $x \in E$.
    Let us introduce fuzzy set $A_{\alpha\alpha}(x)$ with membership function

$$\mu_\alpha(x) = \begin{cases} \alpha & x \in A_\alpha \\ 0 & \text{otherwise} \end{cases} \qquad (8.11)$$

where A_α is an α-cut of fuzzy set $A(x)$. Set $A_{\alpha\alpha}(x)$ for $A(x)$ from Fig. 8.1 is displayed in Fig. 8.4. From (8.10) it is clear that $A(x)$ may be presented in the form

$$A(x) = \bigcup_\alpha A_{\alpha\alpha}(x) \qquad \alpha \in [0,1) \qquad (8.12)$$

which demonstrates the *resolution principle*.

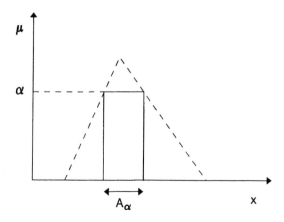

Fig. 8.4 Membership function of set $A_{\alpha\alpha}(x)$

Consider function $y=f(x)$ mapping x from a discourse E_A in discourse E_B of y values. Assume that x are members of fuzzy set $A(x)$ with membership function $\mu_A(x)$. Set $A(x)$ maps in fuzzy set $B(y)$ whose elements are $y=f(x)$ with membership grade $\mu_A(x)$. If f maps more than one element x in the same element $y=f(x)$ in E_B, then the maximum value of corresponding $\mu_A(x)$ has to be attached to y. The rules for mapping outlined define the *extension principle*.

8.2.2
Fuzzy Logic

Let $A(x)$ be a proposition which holds with $\mu_A(x)$. Consider now proposition $B(x)$ which holds with grade $\mu_B(x)$. The proposition

$$C(x) = A(x) \ OR \ B(x) \tag{8.13}$$

holds with

$$\mu_C(x) = \max(\mu_A(x), \mu_B(x)) \tag{8.14}$$

The proposition

$$C(x) = A(x) \ AND \ B(x) \tag{8.15}$$

holds with

$$\mu_C(x) = \min(\mu_A(x), \mu_B(x)) \tag{8.16}$$

Let $A(x)$ be true with $\mu_A(x)$ and $B(y)$ be true with $\mu_B(y)$. Then the statement *if $A(x)$ then $B(y)$* is true with

$$\mu(x,y) = \bigcup_{x,y} \min(\mu_A(x),\ \mu_B(y)) \tag{8.17}$$

The union in (8.17) is applied to all pairs of x and y. $A(x)$ is the *premise* and $B(y)$ is the *conclusion*.

8.3
Dependability Grading

8.3.1
Dependability Evaluation in General

Dependability of a system might be evaluated as, say, high, medium and low. To quantify the dependability more precisely, a dependability grade scale extending from 0 to 100% is introduced [6] and membership functions of grades g in the sets of high, medium and low reliability are defined (Fig. 8.5).

As may be seen, the membership functions overlap partially which corresponds to the nature of assessing linguistic attributes high, medium, low. That is, some grades can be assessed as being both low and medium or being both medium and high to different degrees.

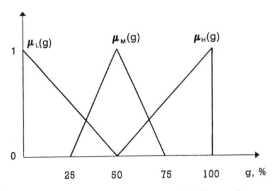

Fig. 8.5 Membership function of reliability grade

In many applications the long-term behavior of a system is of principal concern. For the majority of system customers the severity of service disturbance can be quantified by overall annual duration d and frequency f of service interruptions. By definition, index d is determinable from the steady state unavailability

$$d = U\ T \qquad\qquad (8.18)$$

with T being a year period expressed in appropriate time units, say hours. The general shape of the membership function of dependability index I in the set of unacceptable values is displayed in Fig. 8.6. Parameter a determines the threshold value of index I up to which this index is not considered to be unacceptable to any degree. Index I magnitudes exceeding b are evaluated as unacceptable to the highest degree. Parameters a and b should be specified for both of indices f and d and for each customer individually, depending on their specific requirements and needs.

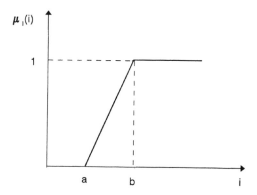

Fig. 8.6 Membership function of reliability index I in the set of unacceptable values

8.3.2
Network Dependability Evaluation

The application of the fuzzy logic based method of dependability grading and evaluation will be demonstrated for a general network [6] which may be representative for a meshed electric power distribution system, telecommunications network, water supply system, etc.

The following *if then* rules can be introduced for evaluating the dependability of a network node, say node j:

Rule I : If f_j and d_j are both unacceptable then the dependability of node j is graded as low.

Rule II : If f_j is unacceptable while d_j is not unacceptable or vice versa then dependability of node j is graded as medium.

Rule III : If both f_j and d_j are not unacceptable then dependability of node j is graded as high.

Index j labels the dependability indices of node j.

By applying the arguments of fuzzy logic theory, the membership function of the dependability of node j in the set of low grades, generated by **Rule I** , equals:

$$\mu_{Lj}(g) = \min(\delta_j , \mu_L(g)) \qquad \forall \, g \qquad\qquad (8.19)$$

Here

$$\delta_j = \min(\mu_{Fj}(f_j), \mu_{Dj}(d_j)) \qquad\qquad (8.20)$$

is the grade with which the premise in **Rule I** holds. Eq. (8.19) is generated by (8.17) where δ_j stands for x and g stands for y. As δ_j is a single value, $\mu_{Lj}(g)$ is a function of g only. Here and further on μ_F and μ_D are the membership functions for interruption frequency and duration in the sets of unacceptable values.

Define

$$\alpha_j \equiv \min(\mu_{Fj}(f_j), \, 1-\mu_{Dj}(d_j))$$
$$\beta_j \equiv \min(1-\mu_{Fj}(f_j), \, \mu_{Dj}(d_j)) \qquad\qquad (8.21)$$

which yield the grade of the truth of the **Rule II** premise clauses.

As we conclude by the same reasoning as before, the membership function of the dependability of node j in the set of medium dependability grades is, with regard to **Rule II**,

$$\mu_{Mj}(g) = \min(\max(\alpha_j , \, \beta_j), \, \mu_M(g)) \qquad \forall \, g \qquad\qquad (8.22)$$

The membership function of the dependability of node j in the set of high grades is, as follows from **Rule III**,

$$\mu_{Hj}(g) = \min(\sigma_j , \, \mu_H(g)) \qquad \forall \, g \qquad\qquad (8.23)$$

with

$$\sigma_j \equiv \min(1-\mu_{Fj}(f_j), \, 1-\mu_{Dj}(d_j)) \qquad\qquad (8.24)$$

The membership function of the dependability of node j in the sets of dependability grades is obtainable by aggregation

$$\mu_j(g) = \max(\mu_{Lj}(g), \, \mu_{Mj}(g), \, \mu_{Hj}(g)) \qquad \forall \, g \qquad\qquad (8.25)$$

Dependability grade g_j of node j is determinable from $\mu_j(g)$ by defuzzification using one of the available methods. We determine g_j as the abscissa of the highest peak magnitude of $\mu_j(g)$. If this peak is flat, which is often the case, g_j is taken to be the peak's lowest abscissa.

Network dependability may be graded using various criteria. The rules defined below provide a reasonable option.

Rule I : If f_j and d_j are not unacceptable for any j then system dependability is graded as high.

Rule II : If f or/and d are unacceptable for only the dependability lowest graded node, say node h, then network dependability is graded as medium.

Rule III : If f or/and d are unacceptable for both of the two dependability lowest graded nodes, say nodes h and k, then network dependability is graded as low.

The grade of network dependability being high is, with regard to **Rule I**,

$$\mu_{NH}(g)=\min[\min_j(1-\mu_{Fj}(f_j)),\min_j(1-\mu_{Dj}(d_j)),\mu_H(g)]$$
$$\forall\ g \tag{8.26}$$

where j applies to all consumers' nodes.

The grade of network dependability being medium equals, as follows from **Rule II**,

$$\mu_{NM}(g)\ =\ \min(\gamma,\ \mu_M(g))\qquad\forall\ g \tag{8.27}$$

with

$$\gamma\ \equiv\ \min[\ \max(\mu_{Fh}(f_h),\ \mu_{Dh}(d_h)),$$
$$\min_i(1-\mu_{Fi}(f_i)),\ \min_i(1-\mu_{Di}(d_i))\ ] \tag{8.28}$$

where i runs over all consumers' nodes except node h.

The grade of network dependability being low is, as follows from **Rule III** ,

$$\mu_{NL}(g)=\min[\ \max(\mu_{Fh}(f_h),\ \mu_{Dh}(d_h)),$$
$$\max(\mu_{Fk}(f_k),\ \mu_{Dk}(d_k)),\ \mu_L(g))\ ]\qquad\forall\ g \tag{8.29}$$

The network dependability grade g_N is determinable by defuzzification, from function

$$\mu_N(g)\ =\ \max(\mu_{NH}(g),\ \mu_{NM}(g),\ \mu_{NL}(g))\qquad\forall\ g \tag{8.30}$$

using the same approach as for nodes.

Example 8.2

Consider the network sample in Fig. 8.7. Node No 1 is the source node while all remaining nodes are customers', sink nodes. All branches and nodes No 2,3,4 are supposed to be unreliable. All branches and all unreliable nodes are presumed to have the same dependability indices, as quoted in Table 8.1.

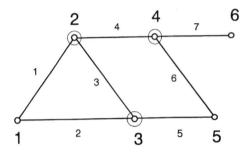

Fig. 8.7 Network sample

Table 8.1. Dependability indices of network elements

Network elements	Failure frequency, fl./yr	Overall annual failure duration, hours
Unreliable nodes	0.2	1
Branches	0.5	10

Table 8.2. Parameters of the membership functions of nodes' dependability indices in the sets of unacceptable values

Parameter	Node				
	2	3	4	5	6
Service interruption duration, hours					
a	0.1	0.1	0.1	0.5	1.0
b	3.0	3.0	3.0	5.0	20.0
Service interruption frequency, fl./year					
a	0.1	0.1	0.1	0.3	0.5
b	1.0	1.0	1.0	1.0	2.0

Table 8.2 lists parameters a and b of the membership functions of customer nodes' dependability indices in the sets of unacceptable values whose general shape is depicted in Fig. 8.6.

Table 8.3 quotes the dependability indices of network nodes calculated by applying the cut set approach, the associated membership grades in the sets of unacceptable values and the dependability grades of nodes obtained by defuzzification.

As may be observed, node No 5 is evaluated as 100% dependable because the values of both its dependability indices are not unacceptable to the highest degree. Node No 6 is graded as 73.9% dependable only, primarily because of the duration of annual interruption which considerably exceeds the threshold value for this node from Table 8.2. The second lowest dependability graded node is node No 4 exposed to comparatively high failure frequency when related to the requirements of the customers associated with this node.

The calculation flow for network dependability evaluation will be presented in more detail, for illustration.

Table 8.3. Calculated dependability indices, unacceptability grades and dependability grades

Node, j	2	3	4	5	6
d_j, h	5.15×10^{-02}	5.15×10^{-2}	0.204	3.46×10^{-2}	10.30
f_j, fl./yr	0.201	0.201	0.291	4.82×10^{-3}	0.704
$\mu_{Dj}(d_j)$	0.0	0.0	3.45×10^{-2}	0.0	0.489
$\mu_{Fj}(f_j)$	0.111	0.111	0.212	0.0	0.136
g_j, %	94.4	94.4	89.0	100.0	73.9

From the data in Table 8.3 it follows that

$$\min_j(1-\mu_{Fj}(f_j)) = \min(0.889, 0.889, 0.788, 1., 0.864) = 0.788$$

$$\min_j(1-\mu_{Dj}(d_j)) = \min(1., 1., 0.965, 1., 0.511) = 0.511$$

With refernce to (8.26) we have

$$\mu_{NH}(g) = \min(0.788, 0.511, \mu_H(g)) = \min(0.511, \mu_H(g)) \qquad \forall\, g$$

Terms in (8.28) are

$$\max(\mu_{F6}(f_6),\ \mu_{D6}(d_6))\ =\ \max(0.136,\ 0.489)\ =\ 0.489$$

$$\min_i(1-\mu_{Fi}(f_i))\ =\ \min(0.889,\ 0.889,\ 0.788,\ 1.)\ =\ 0.788$$

$$\min_i(1-\mu_{Di}(d_i))\ =\ \min(1.,\ 1.,\ 0.965,\ 1.)\ =\ 0.965$$

Having regard to (8.28) we obtain

$$\gamma\ =\ \min(0.489,\ 0.788,\ 0.965)\ =\ 0.489$$

and (8.27) becomes

$$\mu_{NM}(g)\ =\ \min(0.489,\ \mu_M(g))\qquad \forall\ g$$

As stated previously, the two nodes with lowest graded dependability are nodes No 6 and No 4. For $h = 6$ and $k = 4$, (8.29) becomes

$$\mu_{NL}(g)\ =\ \min[\max(0.136,\ 0.489),\ \max(0.212,\ 3.45\cdot10^{-2}),\ \mu_L(g)]$$

$$=\ \min(0.136,\ \mu_L(g))\qquad \forall\ g$$

Membership function $\mu_N(g)$ is obtainable by aggregation, according to (8.25) (Fig. 8.8).

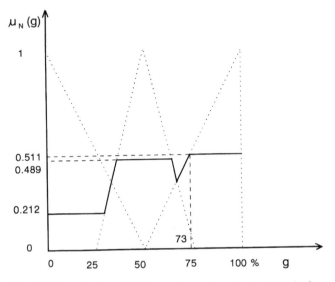

Fig. 8.8 Network dependability grade membership function in the aggregated sets of low, medium and high reliability grades

By defuzzification, using the highest peak lowest abscissa method, the network dependability grade is determined to be $g_N = 73\%$ which could be interpreted as the percentage of satisfying the dependability criteria reflecting consumers' requirements and needs.

□

Problems

1. The capital cost of a system is 5 monetary units. The cost of system failure before 3000 h is 2 units. The sound service time of the system is exponentially distributed with $\lambda = 1$ fl./2000 h. Determine the expected system cost. (Refer to (8.6))

2. Determine the optimal number of systems that should be parallel operating to provide the minimal total cost for the conditions in *Problem 1* (*Example 8.1* and (8.6)).

3. Evaluate the dependability grade of the network in *Example 8.2* if branches 3 and 4 are missing (*Example 8.2*).

References

1. Nahman, J., *Methods for Dependability Analysis of Electric Power Systems* (in Serbian) Naucna Knjiga, Belgrade (1992)
2. Wojczinski, E., Billinton, R., Effects of distribution system reliability index distributions upon interruption cost/reliability worth estimates, *IEEE Trans. Power Appar. & Syst.* **PAS-104** (1985), pp. 3229–3235
3. Kariuki, K.K., Allan, R.N., Palin, A., Hartwright, B., Caley, J., Assessment of customer outage costs due to electricity service interruptions, *Proc. Int. Conf. on Electricity Distribution (CIRED)* (1995), pp. 2.05.1–2.05.6
4. Fischhhoff, B., Lichtenstein, S., Slovic, P., Derby, S.L., Keeney, R.L., *Acceptable Risk*, Cambridge University Press, Cambridge (1981)
5. El-Kady, M., A., Veinberg, M. Y., Risk assessment of grounding hazards due to step and touch potentials near transmission line structures, *IEEE Trans. Power Appar. & Syst.* **PAS-102** (1983), pp. 3080–3087.
6. Nahman, J., Fuzzy logic based network reliability evaluation, *Microelectron. Reliab.*, **37** (1997), pp. 1161–1164.
7. Klir, G., J., Yuan, B., *Fuzzy Sets and Fuzzy Logic*, Prentice Hall Englewood Cliffs, New Jersey (1995)
8. *Electric Power Applications of Fuzzy Systems,* Ed. by El-Hawary, M., IEEE Press, New York (1998)
9. Kaufmann, A., Gupta, M., *Fuzzy Mathematical Models in Engineering and Management Science*, Elsevier Science, Amsterdam (1988)

9 Uncertainties in the Dependability Evaluation

The dependability of a system depends upon various system parameters and indices. Unfortunately, some of these basic data can, sometimes, be only roughly assessed. This is particularly the case when new equipment and constructions are employed whose behavior cannot be predicted with confidence as suitable experience is insufficient or lacking. It is clear that the uncertainty of input parameters reflects on the results of the dependability analysis. In some cases, the uncertainty of these results might be considerable to the extent of making their practical value questionable. In this chapter we will show how to handle the uncertainties and to measure their effects by applying fuzzy numbers [1,4].

9.1
Mathematical Operations with Fuzzy Numbers

9.1.1
Fuzzy Numbers Representation and Evaluation

We consider a normalized fuzzy number $A(x)$ having a single element kernel and finite bounds (Fig. 9.1). Each α-cut yields an interval of possible values of x with lower bound $A_{\alpha l}$ and upper bound $A_{\alpha u}$. For increasing α these bounds become closer to one another tending to a single value as α approaches to 1. This value is the kernel of $A(x)$, by definition. Parameter α may be understood as the *level of uncertainty* with which the fuzzy number bounds are assessed. The closer the bounds the greater is the uncertainty in these bounds.

From Fig. 9.1 it is clear that both the lower and upper bounds of $A(x)$ may be represented as functions of α if the shape of the $\mu_A(x)$ is known, and vice versa. As an illustration, for the triangular fuzzy number displayed in Fig. 9.2 the following expressions for the bounds are obtained

$$A_{\alpha l} = \alpha(\ker(A) - A_{0l})) + A_{0l}$$

$$A_{\alpha u} = -\alpha(A_{0u} - \ker(A)) + A_{0u} \tag{9.1}$$

We leave it to the interested reader to check these expressions.

As may be seen, a triangular fuzzy number is defined by the triple: $(A_{0l}, \ker(A), A_{0u})$.

Obviously, the uncertainty about a number x is the more pronounced the broader are the intervals of guessed values. The *general uncertainty* of x may be quantified using the expression [5]

$$GU = \frac{100}{|\ker(A)|} \int_0^1 (A_{\alpha u} - A_{\alpha l})(1 - \alpha)d\alpha \qquad (9.2)$$

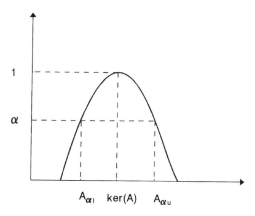

Fig. 9.1 Membership function of fuzzy number $A(x)$

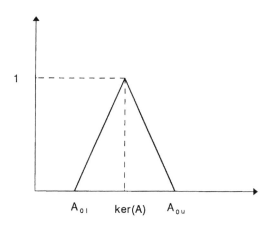

Fig. 9.2 Membership function of triangular fuzzy number

The *lower bound uncertainty* may be defined as

$$LU = \frac{100}{|\ker(A)|} \int_0^1 (\ker(A) - A_{\alpha l})(1 - \alpha)d\alpha \qquad (9.3)$$

and the *upper bound uncertainty*, as

$$UU = \frac{100}{|\ker(A)|} \int_0^1 (A_{\alpha u} - \ker(A))(1-\alpha)d\alpha \qquad (9.4)$$

It is implied that $\ker(A) \neq 0$.

From (9.2) – (9.4) it follows that

$$GU = LU + UU \qquad (9.5)$$

As may be observed from (9.5), GU is defined as the mean interval encompassing guessed x values, expressed as a percentage of $\ker(A)$. The latter coincides usually with the crisp value of x which would be used in the conventional, crisp values analysis. The guessed intervals of x are weighted by the associated levels of guess certainty $(1-\alpha)$. The definitions of lower and upper bound uncertainty grades are self-explanatory. These grades yield the mean expected percentage deviations of lower and upper bounds from the kernel value.

On the basis of the former definitions, the following approximation may be taken as reasonable for engineering decisions made under uncertainty [5]

$$\ker(A) \ (1-\frac{LU}{100}) \leq x \leq \ker(A) \ (1+\frac{UU}{100}) \qquad (9.6)$$

In many practical applications, a single value is sought for adequately representing the fuzzy number $A(x)$. This representative value, say x_c, is obtainable by defuzzification, as already mentioned in Section 8.3. Here, we shall present the center of gravity method which yields

$$x_c = \frac{\displaystyle\int_{A_{0l}}^{A_{0u}} \mu_A(x) x \, dx}{\displaystyle\int_{A_{0l}}^{A_{0u}} \mu_A(x) dx} \qquad (9.7)$$

That is, x_c is the distance of the center of gravity of the area under $\mu_A(x)$ from the ordinate axis.

9.1.2
Functions of Fuzzy Numbers

Let fuzzy number $A(x)$ depend on several fuzzy numbers $A_i(x_i)$, $i=1,...,n$. This fact may be formally written as

$$A(x) = f(A_1(x_1),..., A_n(x_n)) \qquad (9.8)$$

where symbol $f(\ )$ indicates that $A(x)$ is generated from $A_i(x_i)$, $i=1,...,n$.

The associated conventional function is

$$x = f(x_1, ..., x_n) \tag{9.9}$$

Each α-cut of $A(x)$ depends on $f(\ )$ and on the corresponding α-cut of numbers $A_i(x_i)$, $i=1,...,n$. The lower bound of α-cut of $A(x)$ is obtainable as

$$A_{\alpha l} = \min_{x_i \in A_{\alpha i}, \ i=1,...,n}(f(x_1, ..., x_n)) \tag{9.10}$$

That is, the minimum of $f(\ )$ is sought for x_i taking the values from their α-cut. The upper bound of $A(x)$ for uncertainty level α is obtainable as

$$A_{\alpha u} = \max_{x_i \in A_{\alpha i} \ i=1,...,n}(f(x_1, ..., x_n)) \tag{9.11}$$

Suppose that $f(\ )$ is monotonically increasing with $x_i \in A_{\alpha i}$, $i=1,...,n$. Then, by simple reasoning we conclude that

$$A_{\alpha l} = f(A_{\alpha l1}, ..., A_{\alpha ln})$$
$$A_{\alpha u} = f(A_{\alpha u1}, ..., A_{\alpha un}) \tag{9.12}$$

Let $f(\ )$ be monotonically decreasing with $x_i \in A_{\alpha i}$, $i=1,...,n$. Then,

$$A_{\alpha l} = f(A_{\alpha u1}, ..., A_{\alpha un})$$
$$A_{\alpha u} = f(A_{\alpha l1}, ..., A_{\alpha ln}) \tag{9.13}$$

Let $f(\ )$ be monotonically decreasing with $x_i \in A_{\alpha i}$ for $i=1,..., q$ and monotonically increasing with $x_i \in A_{\alpha i}$, for $i=q+1,...,n$. In this case

$$A_{\alpha l} = f(A_{\alpha u1}, ..., A_{\alpha uq}, A_{\alpha lq+1}, ..., A_{\alpha ln})$$
$$A_{\alpha u} = f(A_{\alpha l1}, ..., A_{\alpha lq}, A_{\alpha uq+1}, ..., A_{\alpha un}) \tag{9.14}$$

Example 9.1
Consider the function

$$A(x) = \sum_{i=1}^{n} A_i(x_i)$$

This function is monotonically increasing with all x_i as the partial derivative

$$\frac{\partial f(x_1 ,..., x_n))}{\partial x_i} = 1 > 0 \qquad \forall\ i$$

Consequently, the α-cut bounds are, with regard to (9.12),

$$A_{\alpha l} = \sum_{i=1}^{n} A_{\alpha l i}$$

$$A_{\alpha u} = \sum_{i=1}^{n} A_{\alpha u i}$$

$\square$

Example 9.2
The function

$$A(x) = \prod_{i=1}^{n} A_i(x_i)$$

is monotonically increasing with all x_i *if* $x_i \geq 0$, i.e. if $A_{\alpha l i} \geq 0$, $\forall\ i$. This is clear as partial derivatives of $f(\ )$ with respect to x_i are positive for all i. Note that this would be not the case if any x_i in the corresponding α-cut had negative values. For the case we consider, the bounds of α-cut are, with respect to (9.12),

$$A_{\alpha l} = \prod_{i=1}^{n} A_{\alpha l i}$$

$$A_{\alpha u} = \prod_{i=1}^{n} A_{\alpha u i}$$

$\square$

Example 9.3
The difference of two fuzzy numbers

$$A(x) = A_1(x_1) - A_2(x_2)$$

is monotonically increasing with x_1 and monotonically decreasing with x_2. According to (9.14) we have

$$A_{\alpha l} = A_{\alpha l 1} - A_{\alpha u 2}$$

$$A_{\alpha u} = A_{\alpha u 1} - A_{\alpha l 2}$$

$\square$

Example 9.4
The quotient of two fuzzy numbers

$$A(x) = A_1(x_1) : A_2(x_2)$$

is monotonically increasing with x_1 and monotonically decreasing with x_2 if $A_{al2}>0$. Taking account of (9.14) we have

$$A_{\alpha l} = A_{\alpha l1} : A_{\alpha u2}$$

$$A_{\alpha u} = A_{\alpha u1} : A_{\alpha l2}$$

The quotient exists only if $A_{al2}>0$, $\forall \alpha$. □

9.2
Element Dependability Indices

9.2.1
Reliability

The reliability of an element is determinable as (Section 1.1)

$$R(t) = \exp(-\lambda t) \tag{9.15}$$

If we presume that the element lifetime is exponentially distributed, the failure transition rate λ is a constant parameter. In many cases, however, this parameter is not known with certainty for different reasons and it should be assessed using a limited set of statistical data or expert opinions [1,2]. It is clear that the adoption of a crisp value for λ based upon such uncertain information would generate an only apparently precise result in reliability analysis which might be misleading. The best way to treat λ is to represent it as a fuzzy number. This will attune the results obtained to reality and clearly reveal the effects of input data uncertainties.
    As

$$\frac{\partial R(t)}{\partial \lambda} = -t \exp(-\lambda t) < 0 \qquad \text{for } t > 0 , \tag{9.16}$$

the reliability is a monotonically decreasing function of λ. Hence, with regard to (9.13), the bounds of α-cuts of the reliability are

$$R_{\alpha l}(t) = \exp(-\Lambda_{\alpha u} t)$$

$$R_{\alpha u}(t) = \exp(-\Lambda_{\alpha l} t) \tag{9.17}$$

with $\Lambda_{\alpha l}$ and $\Lambda_{\alpha u}$ denoting the lower and upper bound of fuzzy number $\Lambda(\lambda)$ α-cut.

Example 9.5

Assume that $\Lambda(\lambda)$ has the triangular form: (0.833, 1, 1.25) fl./(1000 h). Reliability $R(100\ h)$ should be evaluated by taking into account the above estimated fuzziness of the failure rate.

According to (9.1) we have

$$\Lambda_{\alpha l} = 0.167\alpha + 0.833 \quad \text{fl./(1000 h)}$$

$$\Lambda_{\alpha u} = -0.25\alpha + 1.25 \quad \text{fl./(1000 h)}$$

Table 9.1 quotes the failure rate and corresponding reliability bounds for several α-cuts, calculated using (9.17). Fig. 9.3 displays the membership function of 100 h reliability in the set of possible values, obtained using the data from Table 9.1.

By applying (9.2) to (9.6) we obtain: UU=0.77%, LU=1.08%, GU=1.85%, x_c=0.9043. As may be seen, the uncertainty in calculating $R(100\ h)$ is moderate. Also, x_c differs very slightly from ker($R(100\ h)$). Generally, such a close proximity reveals only that the corresponding membership function is almost symmetrical with respect to its kernel.

With regard to (9.6) a reasonable guess at $R(100\ h)$ is

$$0.8950 \le R(100\ h) \le 0.9118$$

Table 9.1 α - cuts failure rate and reliability bounds

α	$\Lambda_{\alpha l}$ fl/1000 h	$\Lambda_{\alpha u}$ fl/1000 h	$R_{\alpha l}(100\ h)$	$R_{\alpha u}(100\ h)$
0.0	0.833	1.250	0.8825	0.9201
0.2	0.866	1.200	0.8869	0.9170
0.4	0.900	1.150	0.8914	0.9139
0.6	0.933	1.100	0.8958	0.9109
0.8	0.967	1.050	0.9003	0.9078
	ker(Λ)=1.00	fl/(1000 h)	ker R(100 h)	= 0.9048

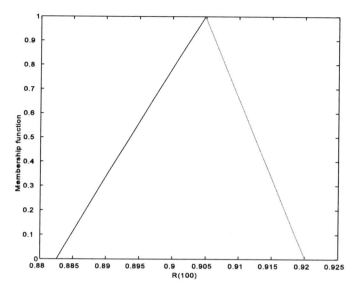

Fig. 9.3 $R(100\ \text{h})$ membership function   □

9.2.2
Steady-State Unavailability

The steady-state unavailability may be expressed as (see (2.20))

$$U = \frac{\lambda d}{1 + \lambda d} \tag{9.18}$$

with d denoting the mean renewal duration. As we observe, index U might be affected by the uncertainty of both the failure transition rate and the mean renewal time.

Partial derivatives of U with respect to λ and d equal

$$\frac{\partial U}{\partial \lambda} = \frac{d}{(1 + \lambda d)^2}$$

$$\tag{9.18}$$

$$\frac{\partial U}{\partial d} = \frac{\lambda}{(1 + \lambda d)^2}$$

As both those derivatives are positive, we have, with regard to (9.12),

$$U_{\alpha l} = \frac{\Lambda_{\alpha l} D_{\alpha l}}{1 + \Lambda_{\alpha l} D_{\alpha l}}$$

(9.19)

$$U_{\alpha u} = \frac{\Lambda_{\alpha u} D_{\alpha u}}{1 + \Lambda_{\alpha u} D_{\alpha u}}$$

where $D_{\alpha l}$ and $D_{\alpha u}$ denote the bounds of the α-cut of the fuzzy renewal time $D(d)$.

Example 9.6

From limited experience it is assessed that the renewal time of the element in *Example 9.5* may be represented by a triangular fuzzy number:(8, 10, 12) h. The uncertainty of the steady-state unavailability of this element should be modeled and estimated.

According to (9.1) we calculate

$$D_{\alpha l} = 2\alpha + 8 \quad \text{h}$$

$$D_{\alpha u} = -2\alpha + 12 \quad \text{h}$$

Table 9.2 quotes the bounds of renewal time and of steady-state unavailability calculated by applying (9.20).

From (9.2) to (9.7) we obtain for this case: $UU=21.30\%$, $LU=14.80\%$, $GU=36.10\%$, $x_c=10.364\times10^{-3}$. These results indicate a remarkable fuzziness of U. There is, also, a noteworthy difference between ker(U) and x_c.

Table 9.2 Bounds of renewal time and steady-state unavailability

α	$D_{\alpha l}$, h	$D_{\alpha u}$, h	$U_{\alpha l}\times10^3$	$U_{\alpha u}\times10^3$
0.0	8.0	12.0	6.620	14.778
0.2	8.4	11.6	7.222	13.729
0.4	8.8	11.2	7.858	12.716
0.6	9.2	10.8	8.511	11.741
0.8	9.6	10.4	9.198	10.802
	ker(D) =	10 h	ker(U) =	9.901×10^{-3}

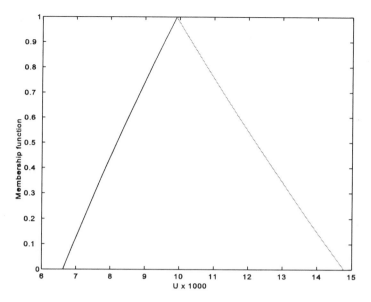

Fig. 9.4 Steady-state unavailability membership function

9.2.3
Time-Specific Unavailability

Consider a two-state element with exponentially distributed residence duration of up and down states. The unavailability of such an element may be expressed as (see 2.16)

$$u(t) = \frac{\lambda d}{1+\lambda d} \left[1-\exp\left(-(\lambda+\frac{1}{d})t \right) \right] \tag{9.21}$$

The partial derivative of $u(t)$ with respect to λ equals

$$\frac{\partial u(t)}{\partial \lambda} = \frac{d}{(1+\lambda d)^2} \left[[1-\exp\left(-(\lambda+\frac{1}{d})t \right)] \right] +$$
$$+ \frac{\lambda dt}{1+\lambda d} \exp\left(-(\lambda+\frac{1}{d})t \right) \tag{9.22}$$

It may be seen that this derivative is positive for all $t>0$.

For the derivative of $u(t)$ with respect to d the following expression is obtained after some elementary manipulations

$$\frac{\partial u(t)}{\partial d} = \frac{\lambda}{(1+\lambda d)^2} \left\{ 1-\left[1+(\lambda+\frac{1}{d})t \right] \exp\left(-(\lambda+\frac{1}{d})t \right) \right\} \tag{9.23}$$

The relationship within the braces is positive for all positive t. Namely, the expression within brackets coincides with the first two terms of the power expansion of $\exp((\lambda+1/d)t)$. As seen, in (9.23) these terms are divided by the exponential function itself which yields a quotient that is less than 1 for all positive t.

Bearing the previous in mind and having regard to (9.12) we may write

$$U(t)_{\alpha l} = \frac{\Lambda_{\alpha l} D_{\alpha l}}{1+\Lambda_{\alpha l} D_{\alpha l}} \left[1-\exp\left(-(\Lambda_{\alpha l}+\frac{1}{D_{\alpha l}})t\right)\right]$$

(9.24)

$$U(t)_{\alpha u} = \frac{\Lambda_{\alpha u} D_{\alpha u}}{1+\Lambda_{\alpha u} D_{\alpha u}} \left[1-\exp\left(-(\Lambda_{\alpha u}+\frac{1}{D_{\alpha u}})t\right)\right]$$

Expressions (9.24) define the time-specific unavailability membership function, making it possible to assess the uncertainty of $u(t)$ caused by uncertain input data, for each t.

9.3
Network Dependability Indices

We imply that networks are dependability coherent which means that the increase of a dependability index of any branch causes an increase of the corresponding network index. These apply to reliability and unreliability, to both the steady-state and time specific availability and unavailability, etc. The former leads to the conclusion that the network reliability is a monotonically increasing function of the reliability of each of its branches. There is the same correlation between network unavailability and the unavailability of network branches. Bearing this fact in mind as well as (9.12), the uncertainty bounds for network reliability α-cut are

$$R_N(t)_{\alpha l} = f(R_1(t)_{\alpha l} ,..., R_n(t)_{\alpha l})$$

(9.25)

$$R_N(t)_{\alpha u} = f(R_1(t)_{\alpha u} ,..., R_n(t)_{\alpha u})$$

with $R_i(t)_{\alpha l}$ and $R_i(t)_{\alpha u}$ being the lower and upper reliability bounds of element i, $i=1,...,n$, for uncertainty level α, i.e. the bounds of the corresponding α-cut. The uncertainty bounds for network time-specific network unavailability are

$$U_N(t)_{\alpha l} = f(U_1(t)_{\alpha l} ,..., U_n(t)_{\alpha l})$$

(9.26)

$$U_N(t)_{\alpha u} = f(U_1(t)_{\alpha u} ,..., U_n(t)_{\alpha u})$$

where $U_i(t)_{\alpha l}$ and $U_i(t)_{\alpha u}$ are the lower and upper unavailability bounds of element i, $i=1,...,n$, for uncertainty level α.

Example 9.7

Let us consider the bridge-type network depicted in Fig. 9.5. We suppose that the branches have the same dependability indices and that these coincide with the indices of the element in *Examples 9.5* and *9.6* . Network reliability for 100 h time interval and network steady-state unavailability are to be assessed.

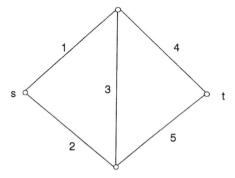

Fig. 9.5 Bridge-type sample network

By applying the minimum path method (Section 4.2) the following expression for the network reliability is derived

$$R_N(t) = 2R(t)^2 + 2R(t)^3 - 5R(t)^4 + 2R(t)^5$$

with $R(t)$ being branch reliability. Network reliability bounds for various α and a fixed t are obtainable by inserting in the above expression the corresponding reliability bounds for the reliability of branches, according to (9.12). By doing so for a series of α values, we obtain the network reliability membership function for the selected time interval t. Fig. 9.6 displays the network membership function calculated for $t=100$ h.

The following results are obtained from the membership function: $UU=0.50\%$, $LU=0.30\%$, $GU=0.80\%$, $x_c=0.9753$, $\ker(R_N(100 \text{ h}))=0.9806$. We can see that the uncertainty in calculating $R_N(100 \text{ h})$ is negligible. Therefore, in calculating $R_N(100$ h$)$, λ could be treated as a crisp quantity being equal to its kernel value. Clearly, $R_N(100 \text{ h})$ is, then, a crisp quantity. Sample network unavailability can be determined by applying the minimum cut method (Section 4.2). By using this approach we deduce that

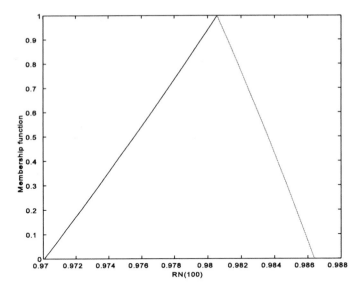

Fig. 9.6 $R_N(100$ h) membership function

$$u_N(t) = 2u(t)^2 + 2u(t)^3 - 5u(t)^4 + 2u(t)^5$$

where $u(t)$ denotes the unavailability of a branch.

For illustration, we shall analyze the steady-state unavailability of the sample network using the data from *Example 9.6*, having regard to (9.11). Fig. 9.7 presents

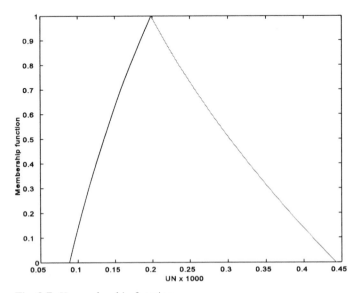

Fig. 9.7 U_N membership function

the calculated U_N membership function. Using (9.2) to (9.7) the following is calculated: $UU=55.80\%$, $LU=30.45\%$, $GU=86.25\%$, $x_c=2.343\times10^{-4}$, $ker(U_N)=1.980\times10^{-4}$. As may be seen, the fuzziness of U_N is considerable. The certainty weighted interval for U_N is, with regard to (9.6)

$$1.377\times10^{-4} \leq U_N \leq 3.688\times10^{-4}$$

This interval is quite broad, indicating that more precise data for renewal duration are needed for a better network unavailability evaluation.

□

9.4
Concluding Remarks

Let us assume that the uncertain input varieties are modeled by normalized, single kernel fuzzy numbers. Then, the output result is obtained as such a fuzzy number too. The benefits offered by the application of the fuzzy modeling are the following:

– The result obtained includes the conventional crisp result as this coincides with the corresponding fuzzy number kernel value.

– Information on the possible values of the output variety of interest due to the uncertainty of some input varieties is obtained. This information is substantially more comprehensive than that obtainable by conventional sensitivity analysis. This latter considers possible deviations of inputs taken one at a time only and in close proximity to a fixed system operating state. The intervals of possible values managed by the fuzzy number approach may be of any size and all uncertain inputs are simultaneously considered.

– The output obtained can be assessed with regard to the grade of the uncertainty by applying (9.2)–(9.4). If this grade is considered too high, the inputs should be reexamined for a more precise quantification.

– Relationship (9.6) yields reasonable, certainty weighted bounds of calculated results reflecting the fuzziness of the input data.

– In preparing the input data the analyst is required to predict the minimum lower and the maximum upper bound for quantity x as well as its perceivable mean value. Usually, these three guesses can be made with a much greater confidence than the prediction of a single representative crisp value for x, as in the conventional approach.

Problems

1. The lifetime of a unit is exponentially distributed with triangular fuzzy failure rate λ: (0.5, 1.0, 2.0) fl./(1000 h). Evaluate the mean time to failure for this element (refer to Section 1.3) .

2. The unit from *Example* 1 has to conduct a mission of uncertain duration. This mission time should be modeled as triangular fuzzy quantity: (100, 120, 180) h. Estimate the probability of mission success (the reliability during the mission time should be assessed for fuzzy failure rate and mission time).

3. The lifetime of a unit is Weibull distributed. The position parameter of the Cdf is zero. Scale and shape parameters are estimated by triangular fuzzy numbers. The estimates are α: (500, 600, 750) h and β: (1.0, 1.3, 1.5). Evaluate the effects upon the unit reliability during 100 h: a) if only α is fuzzy, b) if only β is fuzzy and c) if both α and β are fuzzy (refer to Section 1.3) .

4. A system is constructed by series connection of two identical units. Unit failure rate λ and repair duration d should be dealt with as triangular fuzzy numbers λ: (1.0, 1.5, 2) fl./(1000 h) and d: (7.0, 8.0, 9.0) h. Estimate the steady-state unavailability of the system (refer to Section 4.1).

5. Perform the same analysis as in *Problem* 4 for the parallel connection of units.

References

1. Tanaka, H., Fan, L.T., Lai, F. S., Toguchi, K., Fault-tree analysis by fuzzy probability, *IEEE Trans. Reliab.* **R-32** (1983), pp. 453–457
2. McDonald, J., Noor, S., Forced outage rates of generating units, based on expert evaluations, *IEEE Trans. Reliab.* **R-45** (1996) pp. 138–140
3. Miranda, V., Fuzzy reliability analysis of power systems, *Proc. 12th Power Systems Computation Conference,* Dresden (1996), pp. 558–566
4. Oliveira, A. M., Melo, A.C.G., Pinto, L.M.V.G., The impacts of uncertainties in equipment failure parameters on composite reliability indices, *Proc. 12th Power Systems Computation Conference,* Dresden (1996), pp. 574–580
5. Nahman, J., Peric, D., The quality of supply and the assessment of the technical limits of a distribution network, *Proc. Int. Conf. on Electricity Distribution (CIRED)*, Nice (1999)

Index

Printing: Saladruck, Berlin
Binding: H. Stürtz AG, Würzburg